High-Tech Maneuvers

High-Tech Maneuvers

Industrial Policy Lessons of HDTV

Cynthia A. Beltz

The AEI Press

Publisher for the American Enterprise Institute
WASHINGTON, D.C.

1991

Cynthia A. Beltz is a research associate at the American Enterprise Institute. She received her master's of science in applied economics from the University of California, Santa Cruz.

The author gratefully acknowledges the valuable comments on an early draft of the manuscript from David Richardson, David Mowery, Raymond Vernon, Sylvia Ostry, Theodore Moran, Jeffrey Hart, Philip Webre, and Joseph Donahue. The study also profited from the insights and suggestions of Sven Arndt and Claude Barfield. The views of the study remain the sole responsibility of the author.

Distributed by arrangement with
University Press of America
4720 Boston Way   3 Henrietta Street
Lanham, Md. 20706   London WC2E 8LU England

Library of Congress Cataloging-in-Publication Data

Beltz, Cynthia A.
    High-tech maneuvers : industrial policy lessons of HDTV /
Cynthia A. Beltz.
        p.   cm. — (AEI studies ; 530)
    Includes bibliographical references.
    ISBN 0-8447-3767-4
    1. Television supplies industry—Government policy—United States.
    2. High definition television—Research—Government policy—United
States. I. Title. II. Series.
    HD9696.T463U625   1991
    384.55′2—dc20   91-23567
                                                                      CIP

1 3 5 7 9 10 8 6 4 2

AEI Studies 530

Printed in the United States of America

Contents

What HDTV Threat from Abroad? 79
A Global Perspective 80
Can Government Do the Right Thing? 88

TABLES

FIGURES

Introduction

In October 1989 an open letter sent to President Bush and Congress warned that "HDTV has become a symbol of our national willingness to compete in the strategic industries and technologies of the 1990s, and our failure to develop it could have alarming implications for our economic future." The letter —from business, labor, and congressional leaders— concluded that the importance of HDTV was self-evident because U.S. trading partners were investing hundreds of millions of dollars in it.[1]

More recently the Council on Competitiveness, a private coalition of corporate, labor, and university leaders, issued a report on the competitive state of the United States in technologies spanning nine high-technology industries. The council found that the United States was ahead or holding its own in two-thirds of the ninety-four technologies that are expected to determine U.S. economic progress in the next decade, but it warned that U.S. industries are losing ground in thirty-three. To promote commercial competitiveness, the council urged that the development of generic technologies that cut across a variety of industries be made a national priority.[2]

Both the letter and the council's report reflect the current debate in the United States over the nature of technology policy. At the center of the debate are legitimate concerns about the role of government in a world of imperfectly competitive markets and imperfect bureaucracies. The debate is not about whether technology markets are imperfect or whether the government has a role in encouraging the development of advanced technologies. Or as one international economist stated,

> At issue is not whether there should be American policies to improve the quality and quantity of resources available for production, and the efficiency with which they are used. The question is whether government intervention should be industry specific, or should instead be limited to the financing of education, research, and other activities that support all economic activity, and to the provision of generalized incen-

tives on the basis of which firms may judge where to use the assistance.[3]

The analysis in the following chapters explores the two sides of the industrial policy debate. It starts with the concerns raised by advocates and then examines both the conditions under which these concerns may justify industry promotion and the extent to which these conditions can be expected to hold in the real world for HDTV.[4]

The Strategic Industrial Policy Debate

The notion of the government picking and trying to promote winning industries is not a new one. Attention in recent years has focused in particular on the so-called strategic industries. Although industry groups offer many suggestions for which industries should be selected as strategic, the analytical guidelines are not well defined for picking them and for determining what type and how much assistance they may require from the government.

The growing literature on international trade and imperfectly competitive markets suggests two possibilities for strategic industries. First, some industries may be strategic because they generate super-normal profits or rents (returns in excess of what the resources employed in the industry could generate in other uses).[5] More relevant to recent policy discussion is the second definition of a strategic industry as one that generates external economies or widespread spillovers for other segments of the national economy.[6]

Many economists have doubts, however, about the utility of strategic industry as a practical guideline. Technological spillovers are difficult to measure. The benefits to the national economy are uncertain. Moreover, to determine whether intervention enhances welfare, the potential benefits and costs of the initiative must be examined relative to other uses of public and private resources. In particular the additional long-run benefits from intervention should cover the costs to taxpayers, consumers, and other producers from promotion of the selected industry.

Some argue that the need for empirical evidence and this type of guidance should not be overstated. Those on the side of greater activism contend that the United States cannot afford to remain passive, particularly in the high-paced high-tech sector, where initial advantage is believed to have enduring commercial importance and when other governments are targeting industries such as HDTV for promotion. The time for business as usual has passed, they argue; a U.S. industrial policy must match foreign strategic targeting efforts if the country is to avoid additional losses in its technological capabilities and industrial base. The government, these advocates conclude, specifically should help reallocate private and public resources to favor domestic producers in strategic industries and subsequently to restrain others in an effort to improve national economic performance.

Those who are skeptical about the merits of this case accept the premise that government has a role to play but disagree with industrial policy enthusiasts on the form that this support should take. Critics of industry-specific promotion policies see two main problems with this approach. The first is knowing which industry to pick to generate technological spillovers for the national economy when the pace of technological change is rapid, the nature of change difficult to anticipate, integration of national economies on the rise, and the interdependence of industries increasing as more technologies and components are shared. The second difficulty is determining the necessary form and level of support for the chosen industry under these conditions. Critics also suggest that the long-term competitiveness in the industries selected as strategic and the ability to exploit the opportunities offered by them will likely depend more on general factors, such as the quality of the education system, the cost of capital, and the technological capabilities of a variety of supporting industries, than on industry-specific promotion plans.

Skeptics also believe that industrial policy enthusiasts oversimplify the policy issues by focusing only on imperfections in the marketplace. Critics stress that the real choice is between an imperfect market and an imperfect government. The issue, they argue, is not simply a matter of sufficient information (although this problem should not be minimized, especially in the dynamic high-tech sector)[7] but also fundamentally one of what is politically feasible. They further believe that once the government starts listing the industries eligible for special favors, the consequences are likely to be more imperfect than those of the market. Who, for example, would *not* claim that the future of one's industry is critical to the nation and therefore worthy of private and public favor—especially in a period of economic downturn and rising competition worldwide? Such an incentive to industry to exaggerate may be manageable provided policy makers and their staffs have the time, energy, and a set of accepted procedures to evaluate objectively the merits of a candidate's case. Skeptics question whether these minimum conditions can be met.

HDTV

The arguments on both sides of the debate can be made more concrete by looking at the debate over high-definition television, which some activists have held to be "the model for developing all other key strategic industries of our future."[8] Technically HDTV represents the next generation of television, which promises household consumers a dramatically improved picture, comparable to that of film or twice as sharp as that of conventional sets. HDTV also represents a new era in consumer electronics (televisions, videocassette recorders, and the like) that will use the more sophisticated electronic technologies and components common to advanced digital computers and information systems. This connection is expected to create dramatic new opportunities for the consumer electronics industry and the computer industry to benefit from each other's innovations, such as HDTV and interactive personal computers.

Politically, however, HDTV has come to represent even more. "To miss out on HDTV is to miss out on the 21st century"[9] was a call to action that carried the discussion of HDTV from research labs to the policy front in early 1989, when foreign HDTV development programs were widely publicized in the United States. By mid-1990, one year later, seven HDTV-related bills had been introduced in Congress, ten committees and subcommittees had held hearings related to HDTV, and countless reports and agendas had been circulated by a wide range of interest groups.

A number of the technologies involved with the emerging HDTV industry, such as advanced semiconductor and display technologies, will find applications beyond consumer electronics in a wide range of fields. In addition to computers the fields include medicine, telecommunications, and education. Interest in cross-cutting technologies (or what some have referred to as generic technologies) and their commercial potential brought attention to the HDTV application, but the claims and proposals of activists quickly made the creation of a U.S. television industry and consumer electronics industry the policy issue.

Proponents promote HDTV as the competitiveness barometer for the entire electronics food chain and the nation: emerging as a winner in the HDTV industry reflects the health of U.S. competitiveness as a whole. First pointing to government-supported HDTV programs in Japan and Europe, these advocates argue that the United States is lagging. Then pointing to Zenith as the only major U.S.-owned television manufacturer, they contend that the country is perilously close to falling even further behind. They conclude that without special help from the government to create the industry, HDTV would be the first in a series of industrial dominoes to fall: foreign HDTV promotion programs would make it impossible for the United States to capture the commercial returns and spillovers associated with the

underlying technologies, the market shares of U.S. computer industries and other electronic industries would suffer, and U.S. technological and economic progress could ultimately be curtailed.

To revitalize the consumer electronics industry, advocates argue
that the government's commitment to the HDTV industry would need
to be exceptional. In particular, much of the policy debate has been
shaped by the American Electronics Association largely because its
sweeping proposal aroused emotions on all sides of the larger
industrial policy debate. The association proposed that consumer
electronics receive "national champion"-style treatment and in the
spring of 1989 called for a federal package of $1.35 billion in subsidies
and loan guarantees for the creation of a federally chartered development corporation, the Advanced T.V. Corporation.

The activist case for this type of support makes these key
assumptions:

• A U.S. HDTV industry that makes consumer sets would be an
essential final-product market because it would both cultivate generic
technologies and stimulate sales for the domestic semiconductor
industry.

• HDTV market shares would determine the commercial competitiveness of other American electronic industries and ultimately progress of the national economy.

• Most important, without preferential assistance from the government to match foreign HDTV programs through measures such as
the ATV Corporation, the U.S. industry would not develop—or at a
minimum be delayed—and the nation would lose the returns associated with the development of the underlying technologies.

Lessons from the HDTV Debate

As the policy debate has unfolded over the past two years, the United
States has learned more not only about HDTV but also about the
limitations of a policy process struggling to understand the ever-
changing dimensions of high-tech industries. First, it has become
clear that an economic case cannot be made for intervening to favor
or to speed domestic HDTV production. The strategic importance of
the consumer HDTV industry had been overstated by activists, and
the debate among the experts over the degree of commercial dependence of other industries on the HDTV industry is not yet over.
Consumer HDTV is neither the only path for advancing sophisticated
digital technologies and future information systems nor the all-
important one. The various paths are complicated and interdependent.

Considerable investments and recent advances in the computer and communication industries are, for example, already playing a vital role that should not be discounted.

Accordingly, support for core or generic technologies cannot simply be equated with promotion of the HDTV application. To accept the activist case of consumer HDTV as the link in the electronics food chain that would determine the commercial competitiveness of the other industries and their offspring would misrepresent both the nature of technological change and the digital revolution, which is drawing the various industries closer together.

Second, the threat to national welfare from foreign HDTV research programs has been exaggerated. True, the Europeans and Japanese are ahead in terms of introducing an HDTV system and expensive sets (roughly $34,000 in Japan) to the market. But they are also in the process of backtracking and second-guessing their HDTV investments. The HDTV system that the Japanese have spent two decades developing is based on conventional TV technology, which engineers in the United States have already bypassed in the competition to develop a digital HDTV system. The Europeans face a more difficult problem. Massive research subsidies to develop their analog-based HDTV system HD-MAC (for high-definition multiplex analog components) over the past six years have succeeded in producing a system that few (Europeans included) are willing to adopt. The unforeseen advances in U.S. labs—which surprised even the experts—are further forcing European governments to confront the real possibility that their HDTV plans may have missed the mark and may already be outdated. Ironically, although the United States chose not to heed the advice of activists to underwrite a national HDTV plan, it may still end up with a superior system more adept at communicating with other digital electronics systems in twenty-first century homes and businesses.

Third, neither the innovation of HDTV nor foreign HDTV targeting will fundamentally alter the economics of location in the television set industry for the foreseeable future. More than 60 percent of the color sets sold in the United States are made here largely by foreign-owned manufacturers with domestic value-added around 70 percent. A substantial portion of HDTV sets is also expected to be made here in the future, although this could change. But the costs of transporting the fragile cathode ray tubes used in conventional sets and advances in display technology are likely to play the key role in determining location rather than the nationality of a firm owning the HDTV technology or plant.

In addition HDTV is part of the trend toward the internationalization of research and production activities—a trend that raises new questions about the ability of domestic subsidies to secure a national advantage in particular industries. Two of the major actors in the European HDTV project (Philips and Thomson) have joined forces, for example, with the David Sarnoff Research Center, the former research arm of RCA, to develop an HDTV system for the United States. Alliances developing in other areas related to HDTV suggest that the scope of technological spillovers generated by the broader group of high-definition systems (HDTV is only one type) will be difficult—if not impossible—to confine to Japan, Europe, or the United States.

The HDTV debate has also amply demonstrated the hazards of high-tech politics. Just as multiple technologies are involved with HDTV and the related final-product markets, there are multiple interest groups—each with its own agenda and report documenting its critical importance. Even the most well-intentioned in this environment would have difficulty setting priorities and determining an efficient allocation of national resources based solely on economic criteria.

The United States moreover could not have picked a worse time for boosting the case of industrial policy. The country has long criticized the Europeans for their heavy industrial subsidies, accounting for 3 percent of GDP in some states, as well as their protection of national champions. After investing billions, many in Europe are wondering whether industrial policy has been worth the cost of fragmented markets and publicly padded payrolls: national champions are lagging, and the electronics industry is struggling. With little to show for its investment, the European Community is now reforming the subsidy system and passionately debating what to do next. U.S. moves toward a program of picking and promoting strategic industries would undoubtedly—to the detriment of long-term national interests—undercut the position of those pushing for additional reform and bolster the case of their opponents.

The HDTV debate has demonstrated the inherent complexity, limitations, and hazards of the industrial policy approach for promoting technological competitiveness. Consumer HDTV may become an important industry that will generate positive spillovers for other industries as well as benefit from them. But it is not possible to conclude that private and public resources need to be reallocated to favor domestic HDTV production. If the commercial development of advanced technologies with multiple applications and the location of productive investment in the United States are the policy directives, then the government should first focus on getting the basics right (sensible macroeconomic, education, and infrastructure policies) instead of expending valuable resources trying to predict and to shape the specific form of technological advance. More than anything else, HDTV has illustrated this lesson.

Structure of the Study

The strategic industrial policy debate is not about whether strategic industries can exist but about how the government can support the development of national technological capabilities. Whether government intervention is industry specific or more neutral in orientation will depend on the judgment of policy makers. But HDTV as the model ironically confirms the misgivings of skeptics and weakens the case of activists for industrial policy.

This study is offered as a guide to this model as well as the promises and pitfalls of strategic industry promotion. Two questions were selected to guide the analysis. One, is the emerging HDTV industry strategic in economic terms? Two, is strategic industry promotion necessary? To explore these questions, the study is divided into three major segments. Chapter 1 is an overview of the changes in the global economy that have inspired participants on both sides of the industrial policy debate. Chapter 2 reviews discussions in the trade literature on strategic industries, focusing on interindustry spillovers and the debate between activists and skeptics on the industrial policy prescription. This section concludes with a two-part policy test to investigate the claims and proposals of those vying for the government's strategic industry label. The next three chapters explore how these conditions apply to HDTV as defined by the claims and proposals of industrial policy activists. The study concludes with a section on alternatives to industrial policy for addressing some fundamental concerns relevant, but not unique, to HDTV (chapter 6) and a section on HDTV lessons for future policy debates (chapter 7).

The structure was chosen to provide a base for separating legitimate economic concerns from unfounded policy conclusions. Throughout the study the intent is to inform without overwhelming the reader with economic jargon. Accordingly those well-versed in the general economic arguments may be inclined to move straight to the HDTV discussion. For those well-tutored in the HDTV policy debate but curious about the status of the larger strategic industrial policy debate, chapter 2 attempts to summarize the theoretical debate as it pertains to cross-cutting industries such as HDTV. As is inevitable in a study of this type, several complicated issues have received only cursory consideration; notes have been used somewhat liberally to compensate and to direct those interested toward more detailed references.

•　　　•　　　•

As this study was going to print, the Office of Science and Technology Policy released a report, mandated by Congress, on the technologies it considers essential for long-term national interests. The report represents the views of a key technology policy group and for that reason its misguided analysis on the consumer HDTV industry is particularly troubling. The discussion on high-definition imaging and displays correctly focuses on the potential of the set of technologies rather than on a specific application. But the report suggests that without a domestic base in consumer electronics U.S. producers will find it hard to remain competitive in other electronic sectors and that U.S. prospects for developing HDTV technology are poor because markets are not accurately measuring the industry's potential. With the recent dramatic advances in computer imaging technologies, the spread of international alliances in high-definition systems, the strong growth prospects in the computer industry, and the varied investment strategies and outlooks in the computer and communication industries, the report's conclusions seem unwarranted. Furthermore they are particularly well designed for stimulating renewed calls for strategic industry promotion plans as the White House report virtually echoes the position of HDTV industry activists for the ATV Corporation— which was found to be without merit in the two-year policy debate.

It is also difficult to understand what supports the report's conclusion that the U.S. position in HDTV technology is behind that of Japan and the Europeans. The report refers to the major HDTV program in Europe but fails to acknowledge that the Europeans themselves are questioning whether they have succeeded with their HD-MAC system only in developing another loser that nobody wants to adopt and that is outdated before the market develops. Unexpected U.S. advances in digital transmission have the Japanese scrambling to catch up while scaling back projections of HDTV sales in Japan.

More fundamentally the report reflects the rising pressure on policy makers to suggest that the specific form of technological and economic progress can be predicted and shaped and that the United States should take a more aggressive role in doing just that. If the United States takes this industry-specific path to planning the future and to improving national welfare, then consumers, business, the political system, and chance need to cooperate to make such investments worthwhile. But such cooperation seems unlikely.

High-Tech Maneuvers

1
The Diminished Giant Syndrome and HDTV

Globalization and interdependence are not new concepts, but the extent to which the actions of firms and governments influence, and are influenced by, their foreign counterparts is unprecedented. Another new twist is the major role of technology-driven investments by multinational corporations. The increasingly intricate web of transnational corporate networks underlying these investments and the advances in communications and transportation that make them possible are drawing nations closer together. To keep pace, policy makers worldwide are struggling to understand how these changes are redefining traditional concepts of national technological capabilities. The struggle has revitalized the high-tech industrial policy debate in the United States and sparked the HDTV debate in particular.

Rising Competition and Interdependence

In the postwar period under auspices of the General Agreement on Tariffs and Trade (GATT), the growth of world trade outraced world GNP. New participants emerged, while traditional players such as the United States became more dependent on trade and technology and capital flows from abroad.[1]

In the 1950s and 1960s the United States was relatively self-sufficient, with imports and exports accounting for roughly 7 percent of GNP. By the late 1980s this share had more than doubled, with imports and exports at 16 percent of GNP in 1988. Trade will continue to increase in importance as firms increasingly operate on a global basis to capture scale economies and to exploit market opportunities such as 340 million consumers in an integrated European marketplace. Trade is particularly important for the high-technology, or R&D-intensive, manufacturing industries. Exports accounted for 18 percent of U.S. high-technology product shipments in 1986 versus a 6 percent share in other manufactured products.[2] The composition of trade has also changed, partly in response to corporate production strategies that may encompass multiple nations. Intra-industry trade is, for example, reflected in the overlap in 1988 in the top U.S. export and import product groups, implying trade within the same product group.[3] The United States has also become more dependent on foreign sources of technology and capital. Between 1973 and 1987 the U.S. share of inward foreign direct investment increased from 9.9 percent to 25.2 percent as the global pool of foreign direct investment rapidly expanded.[4]

As a result of these changes, the United States no longer towers above its trading partners—they have caught up or are quickly catching up. Sources of technology have proliferated worldwide. Productivity levels have converged.[5] Per capita income levels in the industrialized countries are now comparable.[6] Technologically sophisticated peers and aggressive rivals have emerged in Europe and the Pacific Rim. The U.S. share of world GDP has decreased from 40.3 percent in 1950 to 21.8 percent in 1980. The U.S. share of the global high-tech market dropped from 51 percent in 1970 to 42 percent in 1986. Underlying this development is increased import penetration: from 5 percent of the U.S. high-tech market to 18 percent over the same period.[7]

Perhaps the most striking and fundamental change facing the United States and its trading partners is the rising importance of multinational corporations. In contrast to previous stages of global integration, these firms are operating more and more on the basis of a worldwide strategy that encompasses both production and research decisions. This may include importing technology from abroad and producing in a variety of national environments to gain access to thriving markets abroad or to improve and to diversify the firm's skill base. Technology-driven investment decisions have led to a wide array of interfirm arrangements including mergers, joint ventures, technology swaps, and research and production consortia that often cross national borders. Strategic alliances are of particular interest. An alliance of this type can be described in broad terms as the joining of two or more firms in an effort to reach a common goal that requires cooperation for some period to pool skills and resources.

The quickening pace of international communications has facilitated this latest stage of global integration by reducing the importance of distance and encouraging tighter links between multinationals and subsequently between nations. The effect has been two-fold. First, policies once considered domestic such as competition and investment policies have become the subject of international disputes. Second, international corporate strategies are forcing nations to reevaluate the meaning of national technological capabilities: it is becoming increasingly difficult to separate "us" from "them" to determine exactly who is domestic. The policy emphasis remains, however, overwhelmingly on nationally owned firms.

Challenges for the 1990s and Beyond

As U.S. dependence on the broader trading arena and the intensity of international competition have increased, the United States has caught what Jagdish Bhagwati, the international economist, has labeled the "diminished giant syndrome."[8] Before the 1980s there was a strong presumption that free trade was the best policy for the United States and that indiscretions abroad could for the most part be tolerated without causing serious harm to U.S. economic welfare. In recent years this position has been tested as the country has become more dependent on foreign technologies and components. The rise of Japan and the four Pacific tigers (Taiwan, Singapore, Hong Kong, and South Korea) has sparked concern in some over whether America will remain a major economic power. Others have responded that the issues have been oversimplified and that debate over who is number one is largely irrelevant: U.S. power is not in doubt because of the size, breadth, and attractiveness of the U.S. political-economic environment. Further the decline in the relative U.S. position was inevitable given the weakened economic state of U.S. trading partners after World War II.[9]

A popular, although unproven and controversial, claim persists nonetheless that the industrial targeting practices of foreign governments have hurt the United States and have speeded its decline relative to its Pacific neighbors. From this perspective, unless the United States responds with similar policies, the nation will have the wrong mix of industrial output and will follow the path of former industrial powers into economic decline.[10]

Positions such as this gained much needed currency from the national insecurities generated by persistent U.S. merchandise trade deficits in the $120 billion range in the late 1980s. Capturing this mood, Robert Samuelson commented in 1987 that the trade deficit has caused the United States to become "increasingly disillusioned and bewildered" by the evolving global economy, provoking a "sense that we're losing control over our economic destiny."[11] Some have pointed to the downturn in the U.S. high-tech trade balance in the mid-1980s as evidence that the United States was losing position not only in traditional industries but also in the technology-intensive industries at the base of its economic strength and competitiveness.[12] Although the trade balance is a poor indicator of national welfare, and

3

macroeconomic factors (national spending and saving patterns) are the principal determinants of a nation's trade balance, the high visibility of Japan's targeting practices combined with its $50 billion bilateral trade surplus with the United States has strengthened the demand for similar U.S. policies.[13]

In this climate Sematech, a consortium of fourteen U.S. semiconductor chip companies, was created in 1987 to improve manufacturing quality and cost performance in the industry. The federal government funds 50 percent of Sematech's annual $200 million budget and has committed its support at least until 1993. In addition the President's National Advisory Committee on Semiconductors has recommended that federal funding be increased and that consumer electronics be promoted as an essential source of downstream demand for semiconductors.

As part of the next stage in clever consumer electronics products, high-definition television represents the latest, but not the last, test in the high-tech industrial policy debate. According to Fred Branfman, Rebuild America, "it is only one of dozens of industries" that are expected to appear before congressional committees in the 1990s seeking help against strong foreign competitors.[14] HDTV thus provides a timely vehicle to examine whether any changes in the marketplace or in the trade literature have improved the economic case for the industrial policy prescription in the 1990s. As background for this discussion, chapter 2 turns to the debate in the literature on strategic industries and the implications for policy.

2
The Strategic Industry Debate

The controversy revolving around U.S. promotion of HDTV is part of the larger, seemingly ageless policy debate over the proper microeconomic responsibilities of the federal government in the marketplace.[1] At issue is whether national support for the development of advanced technologies should be tilted toward efforts to fine-tune the economy through industry-specific promotion programs.

Whereas foreign governments have traditionally been strong supporters and defenders of national champions, the United States has taken more of an ad hoc approach in which targeting of particular industries to advance national economic welfare has been more the exception than the rule.[2] In the 1980s, as American producers faced mounting foreign competition, this hands-off approach has increasingly come under attack. Attention has focused on high-tech industries and strategic industries.[3] Proponents for more activist policies have argued that the nation can no longer afford to remain passive in a world of imperfectly competitive markets, intensive international competition, and foreign industrial targeting: instead, to preserve U.S. technological capabilities and U.S. economic welfare, the nation must match strategic targeting abroad with an American high-tech industrial policy for commercial markets.[4]

This argument rests on three assumptions: (1) industrial policy has a significant net positive impact on national economic growth; (2) benefits generated by foreign industrial policies come at the expense of the United States (zero-sum game); and (3) an American high-tech industrial policy would level the playing field and significantly improve national economic prospects. Clear evidence does not exist to substantiate the first; study of the second shows that this assumption holds only under special circumstances, while the third is open to question particularly when viewed in the context of American political institutions.

To build a base for untangling these and other issues in the policy debate, three fundamental questions need to be asked:

1. Why should the government intervene?
2. What is a strategic industry?
3. Is industrial policy the best instrument available?

The Industrial Policy Prescription

The term "industrial policy" triggers debate even before the subject of strategic industries is introduced. The debate is often oversimplified and phrased in terms that pit industrial policy against a textbook model of comparative advantage based on perfectly competitive markets, Congress against the administration, and national interest versus wider multinational concerns. Some equate industrial policy with centralized economic planning in terms of either aggregate or sector-specific growth. Others use it as shorthand for government attempts to reshape the composition of industrial output and to cultivate winning industries deemed essential for national economic health. Still others define it not in terms of favoring a specific industry but in terms of a process of managing and coordinating the long-term vision of the nation.[5]

The definition used in this study is based on the industrial targeting definition used by the U.S. International Trade Commission: "coordinated government action to direct productive resources to help domestic producers in selected industries become more competitive."[6] The distinctive features of the industrial policy approach are the industry-specific nature of support and the decision to reallocate private and public resources to favor particular firms and industries and subsequently to restrain others in an effort to improve the total performance of the national economy.

By this definition the visible hand of industrial policy is active in many nations. The motivations are manifold, the targeting tools many, and the historical record mixed. Tools to support favored industries include direct subsidies, targeted subsidies through the tax code (such as special depreciation rules), preferential access to credit, restrictions on local content and foreign investment, guaranteed government procurement contracts, and more recently, managed trade such as the 1986 U.S.-Japan semiconductor arrangement.[7]

Why Intervene?

Industrial policy proponents in recent years have focused on high-tech industries as the growth industries of the future that are expected to generate widespread productivity improvements for other segments of the national economy and to boost national productivity and living standards. But why should government intervene?

Intervention on economic grounds is justified in theory as a response to a market failure marked by a deviation in private and public returns that the government can help alleviate.[8] The nation may be willing to pay for some investments, but the return to individual investors is not sufficient to justify the cost, and under investment from society's perspective may result. Research, for example, can often bring benefits to the economy greater than those that accrue to the firms that undertake and pay for it. Others may use such R&D results without cost or reducing that available to others. Firms, however, will consider only the benefits that accrue directly to them in determining their R&D expenditures. The amount of R&D undertaken may thus be suboptimal. Intervention may then be necessary to eliminate the discrepancy between private and public returns. The United States has used tools such as R&D subsidies for basic research to address distortions of this type.

As another reason for intervention, industrial policy advocates have tried to add preferential treatment (as opposed to uniform subsidies designed to encourage research not specific to any one industry) for industries they consider important to the national economic welfare. Their case has received an apparent boost in recent years from developments in the international trade literature on the nature of comparative advantage and increasing returns to scale. The traditional textbook model of trade assumed markets were perfectly competitive, with the pattern of international specialization or comparative advantage[9] explained by intrinsic differences between national economies, with little to be gained from intervention.[10]

A new set of trade models focuses on imperfect competition and increasing returns to scale (declining average cost as output increases) as a better representation of the marketplace for some industries, particularly in the high-tech sector.[11] The increase in trade based on scale economies[12] has heightened interest in these models that suggest that the pattern of trade and comparative advantage can be determined by factors other than inherent differences between economies. Because much of trade is believed to reflect the importance of history,[13] the pattern of specialization may be arbitrary; comparative advantage can be shaped by past economic circumstances whose influence has persisted or grown over time as a result of positive feedback mechanisms in development or production.[14] Comparative advantage and competitiveness are in this respect dynamic rather than static.

If advantage can be shaped rather than dictated, then some argue that the government might be able to intervene and intentionally to reshape the nation's production and trade patterns over time. As evidence that governments can create comparative advantage to the benefit of the nation, industrial policy advocates often point to the targeting of semiconductors and other high-tech industries in Japan and its rapid economic growth. As Harvard Business School professor Bruce Scott has summarized, "It is a new theory of getting and keeping dynamic comparative advantage, a theory that allows a nation to come from behind and overtake a competitor, thus creating an advantage where one did not previously or inherently exist."[15] But what is the significance of the industries in which a nation specializes, whether by accident or intention?

It matters, some answer, because particular industries are more important than others for increasing national income over time.[16] The cause for greater activism, some contend, is thus the prospect of intervening today to cultivate particular industries that will improve the dynamic allocation of resources: national welfare is on a higher growth path than it would be without intervention. (But as is discussed later, this prospect is not a sufficient condition for intervention. Instead important market failures would also need to be evident such that policy makers can determine how much and what form of assistance is needed.) The danger, proponents continue, is that failure to act more aggressively on the part of the United States to specialize in these industries can damage welfare since foreign industrial targeting can affect the price and composition of what the United States produces.

Although it is not often true, the argument is not without merit that industrial targeting abroad—even if it fails to raise income in the home country—could damage U.S. economic welfare. Attention has traditionally focused, for example, on the potential of foreign targeting to affect negatively the terms at which the United States trades exports for imports. Robert Lawrence, a senior fellow at the Brookings Institution, has found, however, that the general decline in U.S. terms of trade over the past decade has had a small impact on U.S. GNP.[17] This decline has also been associated more with structural factors than with foreign industrial policies.[18] Critics of industrial policy have further argued that if the United States is a net importer of products in a targeted industry, it should welcome foreign industrial practices that pay for American consumer benefits. Accordingly, if Japanese video cassette recorders are produced more cheaply than American VCRs, the United States conserves its resources for alternative uses by purchasing their VCRs instead of making American ones. This position is, however, difficult to sustain politically if the Japanese advantage is perceived as unfairly acquired or if the industry in question is perceived as generating critical technological benefits for the national economy over time.[19]

In this context recent attention has focused on the potential negative effect of foreign industrial policies on the pace of domestic technological change and economic growth.[20] Technology-intensive industries are viewed by many as the source of a nation's economic strength.[21] Japan's moves to target the high-end of the technology spectrum in areas such as semiconductors, computers, and aerospace have as a result struck a chord of fear in many Americans.[22] Some fear that unmatched foreign industrial policies will promote production of these industries overseas and damage the competitive interests of the United States.

What Is a Strategic Industry?

Some have suggested that the United States should specialize in those industries with the "greatest technological potential" and capability to generate beneficial paths of technological change for the national economy.[23] An industry would be strategic in this respect because of its potential to introduce, to develop, and then to spread extensively the benefits of new technologies across a wide range of domestic industries so that national productivity is improved. But the process for picking and developing these industries is more complicated than typically assumed.

Some have suggested that to increase national income, resources should be reallocated to favor investment in industries with high value-added per worker. But this condition, Paul Krugman[24] and others have argued, oversimplifies both the strategic industry and the policy questions. High output per worker often reflects high input per worker in terms of physical capital and training (capital-labor ratios). Therefore, if the general rate of investment is not increased, the shifting of investment away from low value-added and toward high value-added industries (that have higher capital-labor and capital-output ratios) could reduce output and employment growth. Charles Schultze of the Brookings Institution has emphasized that a welfare-improving plan of strategic industry promotion would at a minimum require policy makers to determine the propensity of firms to overinvest in low valued-added industries and underinvest in high value-added, the extent to which skilled workers are trapped in low-paying jobs in industries with low value-added per worker, and the extent to which workers from contracting industries will be absorbed into expanding industries.[25]

Another misleading but popular suggestion is simply to choose what other nations have chosen as strategic. Two immediate questions, however, arise. First, what happens if the industry is not a winning industry, but a loser? Second, if the United States contributes to a cycle of targeting matched by countertargeting, will whatever benefits the industry ultimately generates be sacrificed to a spiral of rising protection and trade tensions?

Two sets of minimum conditions for identifying strategic industries suggested by the trade literature may prove a more reasonable guide.[26] An industry may be strategic if it generates supernormal profits or if it generates widespread interindustry spillovers. The first condition, often referred to by economists as strategic trade theory, revolves around relatively new ideas about the government's potential to increase national economic welfare by inducing a change in the production decision of a foreign firm and by shifting a larger share of supernormal profits (price in excess of production resource costs) in a concentrated industry away from this firm toward the domestic firm. The second condition updates traditional ideas about external economies of scale and the prospect of improving the productivity performance in other industries and the national economy.

At issue in both cases is whether the distribution of production across industries and countries matters to national economic welfare. The industrial policy question is whether the nation should try to create competitive advantage by subsidizing and promoting domestic producers in industries that may generate surplus profits or beneficial spillovers for the national economy. In the first case the potential to increase national income through strategic rent-shifting policies has generated considerable debate in the strategic trade literature over the past decade. This study, however, focuses on the case of external economy frequently advanced for promotion of the HDTV industry and other emerging industries.[27]

Technological Spillovers. R&D is an example of a positive spillover (or external economy). This is also referred to as a knowledge or technological spillover. The diffusion of results across firm boundaries may permit a reduction in the production costs necessary for other firms to achieve a given level of technical performance. When these firms do not compensate the innovative firm for benefits that accrue to them, an appropriation problem exists, and underinvestment in R&D from society's perspective may result.

Spillovers of this type are often discussed by economists in terms of untraded interdependencies or effects not mediated by the market. Frequently cited examples include technological linkages between upstream and downstream industries and the relationship between precursor and complementary technologies. Digital computers, for instance, were made possible by advances in microelectronics while advances in integrated circuits were made possible by improvements in scientific instruments and equipment. Another example is the unanticipated advance of scientific knowledge that results from a product's development, as with research for superfast microelectronic switching devices for computers and telecommunications that motivated experiments with superconductivity.[28]

Linkage Spillovers. Spillovers are also often discussed in terms of market-mediated effects between upstream and downstream industries. Increasing returns at one stage of production may translate into increasing returns to society as a whole at the final stage. The classic example of the Swiss watch industry is often used to illustrate how production in one industry can reduce production costs for a linked industry. The machinery used for watch production is characterized by increasing returns to scale in production (average costs decline as machinery output is increased). Increased watch production stimulates additional purchases of machinery and reduces the unit price of the machinery and ultimately of watches; additional production may lead to still lower input costs for the user industry.[29]

Some have referred to cycles of this type as virtuous cycles of interdependence and increasing productivity.[30] They may be relevant to policy because the economies of scale implicit in the watch production example represent potential real savings in resources for the economy. It is important to recognize in policy discussions that scale economies may depend on factors other than the size of the national market and the openness of international markets may thus play a vital role. Japanese computer productivity, for example, may depend on the size of the U.S. market (an international external economy) and on the size of the semiconductor industry (an inter-industry spillover as in the machinery and watch case).[31]

The watch scenario has frequently been extended in high-tech policy discussions to make the case for industrial policy.[32] The case for an individual industry such as HDTV is often described in terms of an entire food chain or system of industries related by technology and input linkages. Instead of generating benefits for a single industry, it is argued, the industry in question is the primary promoter (or driver) of an array of existing and emerging technologies and industries.

The process of determining whether the HDTV industry or other high-tech candidates actually have this type of leverage over other industries is complicated and controversial. Almost all high-tech industries are characterized by some form of increasing returns to scale. Given the increasing convergence of an array of technologies, an increasing number of industries will be linked by technology, input, and customer relationships. Indeed an electronic food chain or system denotes a set of interdependent elements. Determining which activity under these conditions is the strategic one thus requires more than evidence of interdependence.

The spillover argument is also subject to abuse. This was amply demonstrated by one HDTV champion when he justified his call for HDTV industrial policy initiatives on the grounds that "all markets are linked. Lose one market and you lose others. Lose others and pretty soon you lose your entire technological base."[33] But all industries cannot be strategic. Furthermore, binding budget constraints, particularly in recent years, preclude the United States from trying to subsidize and to stimulate domestic production in all industries.

Moreover, although markets may be imperfectly competitive, important gains from specialization and trade are still possible. Nations gain from trade because they can acquire goods from abroad with fewer resources than it would take to produce the goods at home. The high-tech revolution in information systems and imperfect technology markets have not altered this basic insight. Indeed, with increasing returns to scale generated by product differentiation or greater specialization of labor, trade may produce mutually beneficial gains in productivity.[34] David Richardson in a recent empirical study concluded that the gains from increased specialization and freer trade may even be greater when markets are imperfect and comparative costs are not based on resource endowments.[35]

Spillovers and Strategic Industries. Spillovers are also not a sufficient cause for intervention. Appropriation problems and market imperfections clearly exist in the real world, but not every one justifies a policy response. Investment disincentives from spillovers, for example, may be offset by other incentives or market distortions. Such incentives may be to maintain complementary assets or investments in parallel technology development paths such that spillovers generated by the activities of other firms can be effectively exploited.[36] Simply put, spillovers are at least a two-way street. Accordingly policy makers need to look in both directions before determining if and what type of intervention or traffic control is needed. A central issue in the HDTV debate is, for example, why boosting the consumer electronics industry and domestic HDTV production should be considered any more important than competitiveness in other products such as personal computers.

Before intervention can be justified, several additional conditions must therefore be met. First, there must be reasonable evidence that the spillovers generated by the industry in question are more important than spillovers generated by other industries for the nation and that existing market imperfections are so significant that they impede the industry's development in the United States. Second, it must be shown that government intervention can improve the results being generated by the market. Finally, the form of intervention must be carefully structured to avoid doing more harm than good. In this context policy makers should ask whether industrial policy is the best tool available for building strength in the industry in question and for promoting national economic welfare. The following sections consider each of these issues.

Relative importance of an industry. What are some of the conditions that have been used in the literature to narrow the list of spillover industries to the subset of strategic industries that the nation may have a competitive interest in developing? To start, an industry's candidacy for the government's strategic label depends on its ranking relative to other candidates. The list of possibilities includes industries within the same sector such as televisions and computers as well as emerging industries in different disciplines such as the field of biotechnology.

Drawing a line between strategic and nonstrategic industries, however, may ultimately be arbitrary. Instead of one industry, there may be many whose expansion may capture the cost savings associated with the development and production of the common technologies and scale-intensive inputs. How one industry drives another must, therefore, be deciphered. Two related suggestions involve defining the industry's relative importance as a source of demand for the upstream industries and as a force to generate advances in the underlying technologies. The first necessarily revolves around the industry's market size relative to other industries that use the same inputs. The second is inherently more difficult to specify and involves weighing the potential for an industry's performance requirements to advance technologies and the average level of productivity in the economy relative to other emerging applications of the technology.[37]

Country-specific spillovers. An industry's candidacy for the government's strategic industry label also hinges on the geographic diffusion of the spillovers in question. Economists have pointed out that the nation's benefits from advanced technologies or products may not be dependent on the domestic production of a particular high-technology product. Krugman has pointed in particular to Japanese success in consumer electronics initially based on imported semiconductor chips, later Japanese success in chips based on the use of U.S. semiconductor equipment, and the revival of the U.S textile industry in the 1970s based on European capital goods.

Accordingly, there is no reason to presume that spillovers will honor politically defined boundaries. Advances in communication and transportation are tying trading partners closer together; distance is becoming less important; and, with research and production occurring in a variety of nations, industries are becoming more geographically dispersed, further facilitating the spread of technological spillovers. The international diffusion of spillovers is relevant for policy makers to consider in the HDTV debate as well as the broader industrial policy debate because of proponents' claims that industrial programs abroad threaten U.S. welfare and require industrial policy as a defense. But foreign-sponsored spillovers are not a "you win, I lose" proposition: they can spill across national boundaries.[38] U.S.-based firms, for example, may take advantage of foreign-sponsored spillovers through channels such as international technology and production alliances, reverse engineering of product designs, and the purchase of higher-quality inputs.[39]

Before a policy conclusion can be reached, a number of factors need to be considered that will determine whether spillovers are national or international in scope. The type of knowledge produced and the extent of international exchange are two factors of particular importance.

Consider first the spectrum of possibilities for knowledge spillovers. At one end is knowledge specific to a firm, such as experience with a particular manufacturing process, which can be fully appropriated (or internalized) by the firm so that it does not transfer easily to the firm's rivals. At the other end is information that can be appropriated by anyone with the requisite capabilities irrespective of geographic location, such as knowledge embodied in the design of a traded product or published results. Between these two extremes is the knowledge that diffuses beyond the firm's boundaries but remains within national boundaries, such as know-how embedded in a nation's work force.[40]

All three types of knowledge can be found within the semiconductor industry. First, there is detailed production knowledge often associated with learning curves (rising yields and falling average costs as production experience accumulates).[41] A second type is product design that does not raise a question of unique national advantage since diffusion is likely to occur as easily across borders as between firms in the same country. Designs may, for example, be transferred with relative ease to a firm abroad through satellite communication, advanced computers, or engineers involved in an international joint venture. More difficult to identify is the knowledge that does not diffuse easily over long distances and may be embodied in informal communication networks between firms concentrated in a geographic region.

Given this range of possibilities for the geographic nature of knowledge spillovers, there is not a simple rule of thumb to determine whether a technology-intensive industry is strategic. In particular whether a domestic activity creates a special national advantage cannot be reduced to a question of picking innovative or R&D-intensive industries. Rather the source and geographic diffusion of the spillovers believed to be critical for the nation and at risk from foreign programs must also be specified.

These findings are necessary to help define exactly what is strategic for the nation: in which specific industrial activities does the nation need to worry about inadequate domestic investment and the extent of import competition.[42] The semiconductor industry again provides a useful vehicle to discuss the importance of these distinctions. The semiconductor industry is sometimes considered the vital link in the electronics food chain because it generates spillovers for a wide variety of downstream industries. But does national economic growth and competitiveness depend on a strong domestic presence in all the various subindustries and technologies associated with the semiconductor industry such that any foreign targeting of semiconductors poses an economic threat to the United States?

Paul Krugman's simple answer: it is not that simple. In studying the Japanese decision to target random access memories (RAMs) for promotion, Krugman found that they had chosen the area of the semiconductor business least likely to generate country-specific benefits at the expense of the United States. The manufacture of memory chips yields manufacturing knowledge specific to the firm and internationally available product designs but not much in country-specific knowledge that is not appropriable by other firms.[43]

This raises a second point concerning the scope of linkage spillovers. To identify the forces that shape the degree of geographic diffusion and the subsequent strategic importance of these activities for the nation, economists have focused on the extent to which the intermediate good in question is traded.[44]

According to this argument, if the good can be traded, then the scale economies and the related benefits for downstream industries and the national economy depend not on the magnitude of domestic production but on the size of the global market. If total watch demand increases worldwide and if machinery is traded, for example, then the savings in machinery cost from expanded production are not lost to the domestic economy even if the domestic watch industry shrinks. The larger global market permits greater specialization of labor, increased input differentiation, productivity-enhancing spillovers, and globalization of the production process. The rising importance of trade in intermediate products, or the growing volume of intra-industry trade even on a disaggregated basis among developed countries, is consistent with international scale economies of this type.[45]

In this context Krugman has emphasized that linkage spillovers

> are a national competitive concern only if some aspect of nontradability makes them so. In the past such industries as steel, petrochemicals, and capital goods have been portrayed as vital linkage industries; more recently the semiconductor and machine tool industries have been described in the same way. In each case the linkage externality argument is clearly subject to abuse. LDCs [less-developed countries] are free to buy steel and capital goods from more advanced countries (and vice-versa); European countries can make use of Japanese semiconductors and numerically controlled (NC) machine tools. Only where nontradability is significant does a linkage externality arise.[46]

Given this background, two questions immediately arise in the HDTV debate: Is the HDTV industry a vital linkage industry that will create special advantages for the United States? Or are the claims and proposals of those seeking special programs to promote the HDTV industry further evidence of the abuse of the linkage externality argument?[47]

Is Industrial Policy the Best Approach?

A central issue in the strategic industrial policy debate and the HDTV debate in particular is whether government can do any better than the market. It is not sufficient to claim that markets are imperfect because governments are also imperfect. It is also not sufficient to argue that industrial policy can create competitive advantage in a particular industry because it may not be worth the cost. Instead what needs to be shown is that industrial policy will be welfare enhancing: additional benefits from intervention in the long run will cover the costs to taxpayers, consumers, and other producers from promoting the selected industry.

George Eads captured the essence of the policy problem when he noted that "markets are undoubtedly imperfect, but bureaucracies also suffer from imperfections. The relevant comparison therefore is between the results that will be produced by two highly imperfect systems."[48] A welfare-enhancing industrial policy thus requires at a minimum well-established rules and procedures to assess systematically the merits of specific requests for preferential access to capital or other resources and then to administer the appropriate form and level of support.[49]

No Simple Rules. Even when distance matters and spillovers do not readily spread across large geographic distances (that is, spillovers are localized) it does not follow that specialized programs for high-tech industries are necessary. First, the cluster effect commonly associated with Silicon Valley and detailed in Michael Porter's *The Competitive Advantage of Nations* is not unique to high-tech industries. The spillovers generated by geographic concentration of related industrial activities do not set high-tech industries apart from the others.[50] Second, and equally important, localized spillovers, when they exist, indicate the benefits from the concentration of some of the linked industrial activities, but where the cluster should be located either within or across nations is not indicated.[51]

Multiple factors affect the location of successful industries over time including chance, the nature and magnitude of domestic demand, the availability and quality of production inputs, the condition of supplier industries, and the general competitive framework. Porter found that the success of Japanese firms in consumer electronics was based on the complicated interaction of relatively selective buyers, short product cycles, a pool of highly skilled workers, domestic rivalry, and accumulated strengths in the necessary component industries. He found a similar set of intricate relationships at the base of American strengths in computers and complex logic chips.[52]

Many are skeptical about whether the government could intentionally create similar outcomes. In particular, Porter has argued that governments are poorly designed for intentionally creating Silicon Valleys. He has instead emphasized the need for the government to help maintain a competitive and attractive domestic production environment rather than to stray into the area of trying to plan the future by offering specialized incentives to select firms and industries.[53]

The idea of a dynamic comparative advantage challenges policy makers to avoid myopic policies. But how this should be accomplished is not clear. For static notions of comparative advantage, current resources and capabilities relative to other countries provided a guide—specialization in what the nation does relatively well today. For dynamic comparative advantage, the guide is more nebulous since it relies on future American resources and capabilities relative to those of other countries.

To improve national welfare by intentionally creating comparative advantage in particular industries requires at a minimum a long-term plan that indicates where the market is functioning imperfectly today and that anticipates what the allocation of resources across industries should be to generate a dynamically efficient allocation of resources for the United States. The twist is that although the dynamic and unforeseen nature of spillovers provides no guarantee that today's pattern of production is the optimal one for the nation, there is also no guarantee that it is a suboptimal one or more important, that the government could improve the allocation of resources.

The growing trade literature on scale economies and strategic industries has given the economics profession a broader base for studying the intricate nature of economic progress, but policy makers ironically are no better off in terms of knowing which specific industries are strategic or how they should be treated. Perhaps the most telling reflection of the theory's limitation in practice are the comments of Laura Tyson, an advocate for more activist policies in the United States. She conceded that

> Future dynamic gains, precisely because they cannot be signaled and are inherently uncertain, represent a policy gamble. Neither the market nor the policymakers, no matter how clever, can definitely determine whether particular industries are likely to give rise to the most beneficial trajectory of technological change over time. Moreover, even if policymakers, working with private business, scientific, and engineering leaders, could identify the most promising industries and technologies to promote, no guarantee exists that the policies used would be the correct ones.[54]

In summary, then, although strategic industries may exist that will determine national economic growth and health over time, practical guidelines for both identifying them and determining the level and form of necessary assistance do not—even for a panel of experts. As a result, the strategic industry idea—although a powerful suggestion—has not increased support for industrial targeting among economists, and the policy debate remains a divisive one.

The Debate Within. The response of economists and technology experts to the uncertainty associated with strategic industries has been mixed, reflecting fundamental differences in opinions about the relative importance of external economies and the capability of American political institutions to avoid excessive intervention so that national income is increased rather than merely redistributed. While there is some agreement that much of trade is not based on perfectly competitive markets, there is little agreement about what the government should or can do about it: can the government act as a distinctively useful agent in markets on behalf of all its constituents, and if so, should it try to do so?[55]

At the risk of oversimplification, the debate breaks down into skeptic and activist positions.[56] The former starts by asking for evidence that a market failure exists, that it is relatively important for the nation, and that it can be corrected by government intervention to stimulate domestic producers in particular industries. The latter shifts the burden of proof and asks for evidence that a market failure does not exist.

On the activist side researchers at the Berkeley Roundtable on International Economy (BRIE) have argued that externalities in high-tech industries are important and should not be ignored by the government. Tyson has commented that the ability to innovate

> is not easily retained within firms or sectors but is much more easily retained within national borders. Thus, an important externality of market promotion policies in high-technology sectors is the creation of a national pool of R&D talent and the maintenance and strengthening of the human capital necessary to innovate.[57]

When externalities of this type are used to support the case for industrial policy, skeptics have taken issue with the argument for two reasons. The first challenge concerns the form of support chosen to promote human capital development in the United States. In particular skeptics have argued that industrial policy is a poorly designed policy tool if the intention is to increase the quality and quantity of skilled workers. They also question the extent to which the mix of a nation's output determines the availability and growth of skilled workers, instead of vice versa. They suggest that policies designed to increase investment in education and in worker training have a record of generating national returns and therefore are likely to serve national interests better than reliance on uncertain spillovers from favoring producers in particular industries.

Skeptics have also argued that as agents of global integration and increasing interdependence, multinationals have greatly complicated the spillover issue. David Richardson has argued that however large externalities may have been, they are likely getting smaller because of the moves of large multinationals to diversify across sectors and borders to improve their ability to appropriate interindustry spillovers or more formally to internalize externalities. To the extent this occurs, the distinctive role of the government in improving the performance of the market seems to have been at least partially bypassed by the actions of private participants.[58] Additional moves to internalize information flows are suggested by the growth of transnational, corporate alliances in research, production, and marketing.

Economists often attempt to resolve contending policy conclusions through empirical analysis. But it is inherently difficult to quantify the relative importance of technological spillovers without a paper or objective trail of economic transactions. The case studies available also are historical in nature and offer little guidance for policies intended to improve national welfare over time. To what extent is the current commitment of private and public resources to the activity in question inadequate to secure the potential benefits for national welfare? Without clear rules and established procedures to evaluate questions such as these systematically, many economists conclude that it is not possible for the policy process to determine objectively the economic merits of a candidate's case for special attention or to determine the appropriate form and level of assistance. Thus an inherent risk is that the policy will be ad hoc and prone to lengthy debates on individual claims for assistance.

Moreover, the claim that industrial targeting is welfare-enhancing also cannot be substantiated.[59] The ITC found in the early 1980s that the available evidence could not prove that industrial targeting improved national economic welfare in the home country. The ITC added that the evidence of proponents "usually consists of a selection of successful industries in successful countries, assertions that their success is due to targeting, and conclusions that the country's success is due to the targeting of these industries."[60] Conversely cases such as the commercial failure of the supersonic transport suggest that one can win by not picking.[61] Further, although an industry's success and an economy's success may be correlated with industrial targeting, as in Japan, there is no reason to presume that industrial policy has been the main cause of the nation's growth rate.[62]

It would be useful to hold constant all the other factors that may affect growth rates and to explore why they differ across countries to discern what the growth rate would have been without industrial targeting. This is, however, virtually impossible. Rigorous attempts in this area have found that while targeting can alter the composition of what a nation produces, no one has clear empirical proof that industrial policy is welfare-enhancing even in Japan. Marcus Noland found in a recent econometric study that the Japanese industrial policy tools he could measure had a noticeable impact on its trade pattern and generally shifted resources from competitive to less competitive industries. He concluded that "while in some [manufacturing] cases Japanese industrial policy may have successfully targeted industries, welfare-enhancing interventions appear to have been the exception, rather than the rule."[63]

A key limitation of the industrial policy prescription is its industry-specific focus: even if the experts can agree about which industries are strategic, how to build strength in the industries remains a subject of dispute. Strategic industrial policy proponents assume that industry-specific promotion plans are the answer. But a nation's strength in the target industries may be based on broader factors such as average worker productivity, the education system, and strength in a wide variety of other industries. If this is the case, then a policy designed to stimulate selected firms and industries would tend to yield disappointing results at the expense of returns from alternative uses of public and private resources.[64] If the long-run policy objective is to maintain an attractive manufacturing environment to capture spillovers generated by domestic production, then general support for factors of production tied to the United States—such as the physical infrastructure and a skilled labor force—may be a better expenditure of public and private resources. An unfortunate side effect of the industrial policy debate is the shift of attention from the need to promote and to maintain general economic strengths such as the ones both sides agree are necessary.

Yet, as discussed in the introduction to this study, the central issue—rather than a sideline issue—should be the relative benefits of policies that promote general strengths versus those that promote the success of domestic producers in specific industries. How should the government promote national participation in technologically and economically promising industries wherever and whenever they may emerge? The final policy decision will necessarily rest on the judgments of policy makers and their advisers. But the following questions and issues merit the attention of those leaning toward instituting a policy of strategic industry promotion.

Is industrial policy worth the risk? Most realize that promotion programs are not free but require some budget and administrative costs. Economists warn that indirect costs should also not be overlooked. The strategic industrial policy case must show that the benefit side of the ledger from steering resources toward particular industries covers not only the apparent budget costs but also the costs of diverting domestic resources from alternative public and private uses.

Further, the unyielding process of rising technological and economic interdependence in the global economy dramatically complicates the practical problem of promoting domestic production in one industry without damaging the competitiveness of related industries. The 1986 U.S.-Japan Semiconductor Trade Arrangement, for example, raised the price of Japanese chips to U.S. computer firms, damaging their competitive position.[65] Managing a policy of strategic industry promotion thus requires detailed information not only about the industry under consideration but also about the nature and extent of other resource constraints in the economy both today and over time.

Promotion and protection: linked in practice. As another hazard of the promotion of a strategic industry, the initial policy decision could be poorly designed and the costs imposed on other sectors magnified over time, particularly if for political reasons the strategic label cannot be repossessed.[66] This raises the larger issue of the link in practice between promotion and protection. Once an industry is selected as strategic, is it always strategic? What should the limit of government responsibility be, and what is it likely to be? The Sematech experiment illustrates that the pressures for follow-up promotion are real since industries will return and ask for restrictions on foreign activities in the name of safeguarding earlier public and private investments in the industry.[67]

If a strategic industry is one on which national economic strength and competitiveness depend, then the list of industries should be expected to change over time as the national economy changes. History suggests that the list of potential candidates has changed, or perhaps merely grown: textiles and machinery were considered critically important spillover industries in the early 1800s, followed by iron and steel in the late 1800s, automobiles, chemicals, and electrical equipment in the first part of the twentieth century, and more recently various subindustries in electronics.[68] The twenty-first century also promises its share of candidates.

Although an industry may not always be strategic in economic terms for the nation, the accumulated political power of domestic producers in the industry may make it difficult to deny additional support for the industry. Witness the current problems in agriculture in the European Community or the current debate over continued support for textiles and steel in the United States. Furthermore the policy of strategic industry promotion seems predisposed to calls for follow-up promotion and protection because the nation has by definition taken a long-term stake in the domestic growth of a particular industry.

The international trading system. The dynamics of many nations pursuing policies of strategic industry promotion will also affect the long-term costs of this approach both for an individual nation and for the global trading system. This set of factors is perhaps the most critical, and yet the one most often overlooked in the policy debate.

All else equal, each nation acting alone has an incentive to subsidize domestic production in strategic industry candidates if the expected benefits outweigh the national costs of promotion and any necessary follow-up protective measures. Few variables remain unchanged, however, in the real world. National economic opportunities, risks, and decisions are not independent but increasingly intertwined. Considerable economic and political problems may arise as a result if trading partners simultaneously target the same industries for promotion.

This is not an idle concern. One popular suggestion for picking industries in the United States is the crowd-mentality approach to targeting. "The Japanese realize it [HDTV] is important and the Europeans realize it too. And all we're doing is playing ball."[69] The nation should wake up, the argument continues, and stop debating the merits of industrial policy or the possibility of identifying strategic industries: if both the Japanese and the Europeans believe HDTV is strategic, then it must be, and the United States should also target it. But does this make sense? If trading partners target the same industry, important niche markets may be overlooked, potential benefits from specialization may be lost, market overcrowding will occur, government will be pressed to respond by managing trade in the industry with protective measures that fragment markets further, the scale economies everyone is seeking will be reduced, and the global pace of technological progress may ultimately be slowed. It is hard to imagine anyone winning in this context.

Some will argue that this scenario is extreme, but it amply illustrates that the strategic industrial policy approach is not just a challenge to American policy makers but a challenge to the international trading system. On this point activists are conspicuously silent, leaving unanswered how problems of excessive duplication of R&D, congestion, and rising trade tensions can be avoided or addressed if— as seems likely—each nation pursues the same list of strategic industries.[70]

Managing Strategic Industry Lists without Rules. Without firm rules and the risk of costly mistakes considerable for both the nation and the international trading system, few economists are optimistic about whether a U.S.-style strategic industrial policy can be welfare-enhancing. Some, however, remain optimistic and have suggested that the policy gamble be managed by keeping the list of candidates short. But how will the list be kept short in practice? And without detailed policy guidelines how can the nation be sure that the composition of the list will be determined by economic criteria?

The list-making industry is already working overtime. Lists of critical technologies are particularly popular. The stated reason for the lists initially seems straightforward: the social benefits of some technologies are clearly evident but exceed perceived private marginal benefit so the investment that is undertaken today is inadequate to generate the socially optimal result. Several members of BRIE, for example, suggested in 1989 that only those technologies that promise to "radically transform products and production processes of a wide range of sectors" should be chosen as strategic. They emphasized new materials, microelectronics technologies, superconductivity, and biotechnologies. Under the strategic label they suggest that intervention to subsidize the fixed costs of development and to help create initial markets could be justified.[71] More recently the lists of critical technologies circulating through trade associations, the Defense and Commerce Departments, and the administration include ten to thirty critical technologies.[72] Digital imaging technologies, which underlie the HDTV and computer industries, are among the additions to the list.

The list makers these days are, however, extremely vague about the nature and amount of government support they believe is necessary to advance U.S. interests in successful commercialization of the technologies. To avoid the stigma attached to the label of industrial policy, most also explicitly state that picking winners and losers is not part of their policy agenda. They suggest instead that the United States should make the support of critical technologies that cut across industry borders a national priority. Around the corner are more explicit and ambitious industry-specific plans. In the spring of 1990 Rep. Norman Y. Mineta (D, Calif.) called for the secretary of commerce to devise a list of critical American industries that "will provide the bulk of economic opportunity and economic growth" in the future.[73] Deciding which industry to include and exclude, however, sounds suspiciously like picking winners and losers.

Pressure will also undoubtedly continue to build to tilt broad concerns over U.S. technological and commercial capabilities toward preferential treatment for selected industries as competition worldwide continues to increase and perception of unfair foreign trade practices persists. Two issues in this debate merit particular attention.

Advanced technologies and strategic industry claims. In the debate over the nature of U.S. technology policy, the strategic industrial policy case should not be confused with the one for advanced technologies, which may have multiple applications. The case for supporting the development of these technologies in the precompetitive stage is fundamentally different from the one for strategic industry promotion. Support for the former is by definition broad in orientation, while the other is selective. Support for generic technologies should not try to pick a development path, while it is the objective of the latter to steer resources toward particular industries and applications. One defines a limit for government support, while it is in effect left open-ended in the strategic industry case.

Imperfect political markets. The issue of strategic industry promotion is not only a matter of foresight but also inherently one of compatible incentives and institutions. The length and substance of a strategic industry list, for example, are likely to depend more on the nature of nation's political institutions than on the ideas of economists and technology experts about strategic industries.

Some in favor of more activist industrial policies have suggested that the United States should not be deterred by the policy gamble associated with picking strategic industries, especially since the European and Japanese apparently do not have such reservations.[74] Differences in institutions and capabilities across countries preclude, however, this simple comparison. Leaving aside the information requirements and the questionable record of success in Europe, calls for more industrial targeting in the United States need to be evaluated in the context of the special nature of American institutions.

The dynamics of the American policy process, for example, seem likely to be at odds with those of strategic industries. The strategic industrial policy approach requires difficult choices about the distribution of private and public resources, which conflicts with the incentives of elected officials to avoid decisions that give the appearance of favoring one group over another. As a result the politics of building a constituency for industry-specific assistance may spread activities across the United States such that the potential for reaping the technological benefits from the geographic concentration of related industrial activities (local spillovers) is curtailed rather than enhanced by intervention.[75]

The issue is not simply whether the nation can pick industries to promote commercial interests. Clearly it can try to do so. Rather the issue is whether the United States can pick only those candidates that merit selection on the basis of economic criteria and whether assistance can be structured such that national welfare is improved rather than curtailed. Another important factor is whether assistance to the industry can be discontinued if the policy gamble does not pay off for the nation. There is no reason to presume that a strategic industry list could be kept short in practice. Indeed the reverse seems more likely. The preestablished economic conditions for identifying strategic industries are vague, open to interpretation, and therefore likely to encourage strategic industry claims without providing a defensible rationale for resisting pressures to attach the strategic label to a variety of activities. Moreover, since interindustry spillovers are not uncommon, practically any industry could marshal a case for strategic treatment.

Skeptics have argued in this context that the American political process is not well equipped for the task of identifying or determining how to assist potentially strategic industries. Anne Krueger has pointed out that relying on the political process to select strategic industries without a set of clearly defined rules and procedures invites capture by special interest groups. Or, in the words of George Eads, "if care is not taken, 'externalities' will be discovered primarily where it is politically advantageous to discover them."[76]

To heed his advice, it is incumbent upon those who believe that the United States needs to take a more activist stance to explain how policy makers will know in practice which industries or subcomponents should be included on the list and how much assistance they require. Also, how can they exclude those seeking to jump on the bandwagon and push off those that no longer merit preferential treatment? Is it possible to insulate the selection process from political pressures? Activists need to show specifically how the abuse of interindustry spillover arguments for strategic industry promotion will be avoided or minimized. At a minimum, flexibility would have to be designed into list management so that preferential treatment could be discontinued if the policy gamble appears unlikely to pay off for the nation. Otherwise, without mechanisms of this type and without rules, the political-economic cycles engendered by strategic industrial policy seem more likely to be vicious than virtuous.

Policy Implications for the 1990s

The friction between the concept of a strategic industry and its application can be summarized by four simple statements: If we do not change paths, we may lose (dynamic comparative advantage). Yet we cannot know in advance where we are going (the intricate nature of economic progress). It is unclear how we will get there or how we will know once we have arrived (no simple rules or accepted procedures). And if we follow each other, we may all get lost (crowd mentality).

The literature on strategic industries has opened the door for greater activism. It is incumbent upon those who want to draw the U.S. government through this door to explain the necessity of favoring a particular industry. An equally important second step involves determining the distinctive role of government in redressing the underlying concern so that it is a better partner not only for the industry in question but ultimately for the nation.

To chart a path through the inevitable maze of competing agendas, arguments, and counterarguments, American policy makers should reach for two important tools. One is a global perspective to assess the geographic nature of the spillovers in question. As a start, this requires documenting the degree to which foreign actors are helping to develop the domestic industry and U.S.-based firms are participating in foreign markets, and the extent to which international research or production alliances may generate benefits for the United States. The second tool is a ruler to measure as accurately as possible the relative importance of the industry.[77] Policy makers should also recognize that the mixing of strategic industry ideas with the policy process has opened a rather large can of worms, which defy even their organizer's attempts to manage. Krugman has suggested that strategic industrial policy be tried in the 1990s but not because of any newly found faith in the underlying economics and the ability of the political system to support the most promising industries. Rather his intent is to avoid a worse outcome by buying off with targeted subsidies those interest groups promoting the more damaging approach of managed trade.[78] Other professionals who seek to advise the government on the policy implications of strategic industries do not recognize this limit. Instead in their version, managed trade and industrial policy are integrally linked in a strategy to promote high-technology industries.[79] Still others are not willing to exclude traditional manufacturing sectors such as textiles from the strategic industry list.[80] Moreover, little evidence suggests that the bribe of subsidies would appease interest groups from returning for bilateral or multilateral quota arrangements.[81]

This is not to suggest that the United States should minimize the potential economic importance of emerging, innovative industries. Instead it is to underscore that industrial policy is a deceptively simple approach to national competitiveness. In practice on a day-to-day basis the policy requires overcoming a complex course of economic, technical, and political hurdles that may not even be worth the effort.

Industry-led Targeting. In an apparent effort to avoid the sting of critics, enthusiasts of industrial policy have in recent years advertised an industry-led model as an alternative to the government selection of winners and losers. The apparent hope is that industry-led targeting will reduce the range of error and likelihood of negative welfare effects. In this model individual firms will take the initiative and join together to develop domestically owned and controlled industries. The distinctive role of government is to assure the firms at the outset that they will be backed by full government support comparable to what foreign companies receive from their governments. This support might include subsidies, low-cost government loans, matching grants to consortiums, and guaranteed buys of early products produced in the domestic industry.[82]

But is the industry-led approach to targeting an improvement or another label without distinction for industrial policy? As with industrial policy, an incentive to exaggerate on the part of industry is inherent in the new model. In addition, if a domestic industry is still emerging or does not yet exist, then who leads? Who represents the industry? If not government, then who will develop an initiative for this industry-government partnership? In the case of the HDTV industry, the American Electronics Association is the designated driver to develop an industry-led HDTV initiative to help the government promote the creation of the domestic industry.[83] As a result the HDTV initiative that emerged in the spring of 1989 provides a useful vehicle to explore the model activists have advertised as the new industrial strategy to develop strategic industries and to advance America's economic interests. Before this discussion several general cautionary comments about the industry-led model are in order.

Part of the enthusiasm for trying an industry-led approach in the United States stems from the public-private initiatives being tried in Europe.[84] The European objective is an independent European technological capability. Toward this end the European Community invests an estimated $2.1 billion annually in large-scale collaborative multi-country programs such as Esprit, launched in 1983 with more than 1,200 firms involved in information technology projects, and more recently Jessi, to advance microelectronics. Improving the performance of European-owned firms is the primary focus of both.

The Europeans are, however, encountering some troubling imperfections in their version of the industry-led model. Jessi's plan for advancing national technological capability has been fundamentally challenged by the decision of founding member N. V. Philips, a Dutch electronics firm, to exit from a key advanced chip project. By contrast, the most profitable European computer firm, International Computer Ltd. (ICL), was kicked out of three Jessi projects when Fujitsu purchased an 80 percent share in the company. Together the moves undercut a central premise behind the programs, namely, that nationally owned businesses can act as proxies for improving national technological capabilities. A highly placed EC official told the *Financial Times* in July 1990, "I am totally skeptical about the purpose of these collaborative programs. The companies have not carried out their part of the bargain. Instead of using Community money to speed up product innovation and improve productivity, they have accumulated war chests to buy each other up."[85]

Exploring Strategic Industries and HDTV. The European experience confirms that industrial policy measures (even those led by industry) can produce disappointing results. In addition to this background, policy makers should give due consideration to each of the following issues before tilting in the direction of strategic industry promotion: has reasonable evidence been presented that confirms the industry's claim that domestic participation is necessary for national future economic growth, that appropriation problems or other imperfections in private markets impede the emergence of this industry, and that the positive spillovers are geographically constrained such that national competitive interests are considered threatened by promotion policies abroad? In practical terms this last point is perhaps the most pertinent since much of the alarm behind the call for more aggressive U.S. industrial policies is based on a fear that benefits generated abroad come at our expense.

The arguments on both sides of the industrial policy debate can be made more concrete by looking at the recent debate about the emerging high-definition television industry and U.S. competitiveness. The following chapters examine whether the conditions identified above are likely to hold in practice for HDTV and how American policy makers can operate to improve the results generated by the market and industrial policies abroad. Two questions frame the discussion:

1. Is HDTV strategic? How important is the HDTV end-market relative to other electronic markets for generating technological and linkage spillovers for the nation? Are the spillovers likely to be limited to Japan, Europe, the United States? Or are they expected to be international in scope?

29

2. Is promotion of HDTV as a strategic industry necessary for U.S. participation? Is it the best tool available to encourage the development of HDTV as well as other high-definition systems in the United States? Is the industry-led HDTV model compatible with U.S. political institutions?[86]

For background on these issues, chapter 3 turns to the structure of the U.S. consumer electronics market.

3

Television and the Consumer Electronics Market

Much of the alarm in the HDTV debate has stemmed from the concern that the United States does not have a television industry and from the assumption that public assistance is needed to promote the creation of a domestic HDTV industry to revitalize the consumer electronics industry. Before the discussion in chapters 4 and 5 of the main assumptions that drive this policy conclusion, the structure and main trends in the industry merit attention.

Consumer electronics equipment is only one part of the total global market. In 1988, electronic equipment, according to Dataquest, a market research firm, this market was $674 billion, of which roughly 20 percent was consumer electronic equipment. The three major consumer electronic markets are the United States, Europe, and Japan.

The United States and Europe are the two largest markets for consumer electronics equipment and together accounted for slightly more than 50 percent of worldwide television demand in 1988. Both countries are net importers of consumer electronic equipment, with Mexico, Japan, and Southeast Asia the main suppliers.[1] Japan has focused on consumer electronics since the 1950s and accounted for 36 percent of global production of consumer electronics in 1988, followed by Western Europe at 15 percent and North America at 13 percent.

In the early 1980s world production of color television was dominated by Philips, Matsushita, Sony, and Toshiba.[2] Sony has participated in HDTV since the late 1970s and is expected to be a major player both on the household consumer and production side of the business. This reflects in part the rising pressure on Japanese firms to promote the next generation of television products such as high-definition television as they are forced to respond to the competition from newly industrializing countries (NICs) that are in lower-wage East Asia (Korea, Taiwan, and Singapore).

U.S. consumer electronics sales were a $30 billion market in 1987, or roughly 12 percent of the $248 billion U.S. electronics market.

TABLE 3–1

FACTORY SALES OF CONSUMER ELECTRONICS PRODUCTS, INCLUDING
IMPORTS, IN THE UNITED STATES, 1977–1988
(millions of dollars)

Year	Mono TVs	Color TVs	Projection TVs	VCRs	Video Discs	Audio Systems	Audio Components
1977	530	3,289		180		606	1,275
1978	549	3,674		326		748	1,143
1979	561	3,685		389		748	1,178
1980	588	4,210		621		809	1,424
1981	505	4,349	287	1,127	55	720	1,363
1982	507	4,253	236	1,303	54	573	1,181
1983	465	5,002	268	2,162	81	630	1,268
1984	419	5,538	385	3,585	45	976	913
1985	309	5,562	488	4,738	23	1,372	1,132
1986	328	6,024	529	5,258	26	1,370	1,358
1987	287	6,271	527	5,093	30	1,048	1,715
1988[a]	200	6,530	525	5,055	40	1,140	1,800

a. estimate.
SOURCE: Electronics Industry Association, *1987 Electronic Market Data Book*, p. 6;
and *Consumer Electronics U.S. Sales* (Washington, D.C.: EIA, January 1989).

Audio products, video recording, and TV sets are the three major segments of the consumer electronics industry. Televisions and videocassette recorders together accounted for 40 percent of consumer electronics sales with the $6.3 billion color television market as the largest single submarket.[3] (See table 3–1.) Nearly 98 percent of U.S. households own a TV, and most new sales are extra sets or replacement sets.

Across from the consumer side of the TV business, the television production equipment market is dominated by the American firm Ampex and by Sony. U.S. shipments of broadcast and other video production equipment were $2 billion in 1989. A number of specialized markets are characterized by small firms and rapid innovation.[4]

A Global Industry

Of particular relevance to the HDTV debate, the television receiver industry is part of the trend toward global industries that obtain technology, capital, and components from a variety of sources. Japanese firms, for example, make televisions in the United Kingdom and do advanced television research in Germany.

In the United States, foreign-owned firms also play a key role. The Commerce Department estimates that 60 percent of all color televisions sold in the United States are manufactured here with foreign-owned factories producing more than 80 percent of the televisions made here.[5] Foreign TV manufacturers became particularly active in the late 1970s through large-scale acquisitions and the establishment of subsidiaries in the United States.[6] The transfer of design and manufacturing techniques from Japanese firms helped to improve U.S. productivity in some cases as suggested by the 30 percent productivity improvement after Matsushita acquired Motorola's factory in Illinois.[7] At the same time U.S.-owned firms sought to counter the design and production advantages of the Japanese subsidiaries by shifting production offshore in search of lower labor costs. By the end of the 1980s, foreign firms were thus more American in their production activities than many American-owned firms.

There is some dispute over what caused the exit of U.S.-owned firms from consumer electronics.[8] Most accept, however, that the change in the pattern of ownership (from domestic to foreign) did not have a significant independent impact on the structure of the television industry: despite the change several foreign-owned picture tube plants came on line in the late 1980s, and the number of firms in the U.S. television receiver industry increased from fourteen in 1979 to nineteen in 1989.[9] (See table 3–2.) The market is as a result competitive with multiple companies vying for market share or for control of new technologies and products.

The content of television sets produced in the United States includes components and subassemblies produced abroad that account for roughly 20 percent of the factory value of the sets. Given the role of international specialization, the Advisory Committee on Advanced Television Service of the Federal Communications Commission concluded that even if all U.S. television receivers were produced by U.S.-owned companies, domestic content would likely remain essentially unchanged at 80 percent. Industry analysts have also suggested that most of the next generation of television sets, HDTVs, will be produced in the United States with a domestic content ratio remaining between 70 and 80 percent.[10] Domestic content tends to increase with set size and is primarily associated with the use of cathode ray tubes (CRTs), which are manufactured in the United States because of the bulk and fragility of the picture tubes. If and when large flat panels become a viable option for HDTV, the domestic content ratio may change.

Recently the U.S. television market has been dominated not by Japanese firms but by Thomson and Philips. Together they control

TABLE 3-2
PRODUCTION OF TV SETS IN THE UNITED STATES, 1988

Company	Type of Plant	Location	Employees	Annual Production
Bang & Olufsen	Assembly	Compton, Calif.	n.a.	n.a.
Goldstar	Total[a]	Huntsville, Ala.	400	1,000,000
Harvey Industries	Assembly	Athens, Tex.	900	600,000
Hitachi	Total	Anaheim, Calif.	900	360,000
JVC	Total	Elmwood Park, N.J.	100	480,000
Matsushita	Assembly	Franklin Park, Ill.	800	1,000,000
American Kotobuki (Matsushita)	Assembly	Vancouver, Wash.	200	n.a.
Mitsubishi	Assembly	Santa Ana, Calif.	550	400,000
Mitsubishi	Total/Assembly	Braselton, Ga.	300	285,000
NEC	Assembly	McDonough, Ga.	400	240,000
Orion	Assembly	Princeton, Ind.	250	n.a.
Philips[b]	Total	Greenville, Tenn.	3,200	2,000,000
Philips	Tubes	Ottawa, Ohio; Emporium, Pa.	2,300	3,000,000
Philips	Components	Jefferson City, Tenn.	1,000	700,000

Company	Type	Location		
Samsung	Total	Saddlebrook, N.J.	250	1,000,000
Sanyo	Assembly	Forrest City, Ariz.	400	1,000,000
Sharp	Assembly	Memphis, Tenn.	770	1,100,000
Sony	Total	San Diego, Calif.	1,500	1,000,000
Tatung	Assembly	Long Beach, Calif.	130	17,500
Thomson [c]	Total	Bloomington, Ind.	1,766	3,000,000 +
Thomson	Components	Indianapolis, Ind.	1,604	n.a.
Thomson	Tubes	Scranton, Pa.	1,242	n.a.
Thomson	Tubes	Marion, Ind.	1,982	n.a.
Toshiba	Assembly	Lebanon, Tenn.	600	900,000
Toshiba	Tubes	Horseheads, N.Y.	1,000	1,000,000
Zenith	Total	Springfield, Mo.	2,500	n.a.

NOTES:

a. Total manufacturing involves more than the assembling of knocked-down kits.

b. Philips is planning to build a $100 million plant in Michigan to manufacture picture tubes including HDTV.

c. Thomson is planning an expansion of Indiana plant for the production of large-screen and 16 × 9 picture tubes.

SOURCE: Electronics Industries Association, HDTV Information Center.

close to 40 percent of U.S. sales and operate ten plants in the United States, which employ almost 48 percent of American workers in the TV industry. Zenith, the only major U.S.-owned firm remaining in the industry, maintains one television and one picture tube plant in the United States, accounts for 16 percent of the American color TV production work force, and relies on Mexican facilities for 60 percent of all the televisions it sells in the United States.[11] Both Thomson and Philips perform most, if not all, of their TV research for the U.S. market in the United States, and both are heavily involved with American and European HDTV research programs.

One outcome of this research is a HDTV research consortium involving Philips, Thomson, NBC, and the David Sarnoff Research Center which links together two of the leading video research centers in the United States.[12] Zenith also invests in consumer electronics research in the United States and is a major contender with AT&T in the competition to develop a U.S. HDTV transmission system. Sony, Matsushita, and NEC also have consumer electronics research and development centers in the United States, but as of late 1989 only Sony had announced plans to fund HDTV research ($20 million) here in their new research center in San Jose.[13] These producers, all operating in the United States, are expected to have the edge in a number of the technologies that underlie successful HDTV development.

As part of their global competitive strategy, all the major U.S. players have Mexican plants to serve their operations in the United States. These plants specialize in low-end color TV products, where price competition is intense and production in the United States is not affordable. Mexico is also the largest supplier of sets to the U.S. market and reportedly exports twice as many sets as Japan, Taiwan, and South Korea combined.

This role has placed Mexico in the middle of a dumping controversy over the use of some Asian tubes in Mexican sets that are then sold in the U.S. market. The charge is circumventing a U.S. government order intended to assure "fair trade" in television sets by dumping Asian tubes in Mexico used in sets sold in the U.S. market. Zenith, which has made dumping accusations, also uses Asian tubes in some sets it imports from Mexico. Foreign-owned firms that operate in the United States and use Asian tubes in their operations are on both sides of the debate. This case amply illustrates the difficult and often frustrating task of determining who is the victim or who should represent the industry in an era of internationalized markets and flexible sourcing decisions.

The wave of the future in consumer electronics is expected to include a variety of computer-based products. The market for whatever products emerge is also expected to be competitive and international in scope since few, if any, markets are likely to be large enough to offset the fixed costs of development and production. For the emerging HDTV industry in particular, the structure and competitive nature of the U.S. consumer electronics market suggest two points relevant to public policy discussions. First, calls to revive the industry based on domestic ownership are shortsighted and underestimate the prospective costs of this route since Zenith controlled only 12 percent of color TV sales in the United States in 1990.[14] New entrants are likely to be discouraged by low profit margins and intense competition, which should continue to characterize the U.S. market.

Second and most important, if experience is any indication, even a domestically owned TV industry will not necessarily secure HDTV production in the United States. HDTV is unlikely to change the economics of location in the television set industry. Furthermore the increasing integration of Mexico into the North American market and the increasing importance of Pacific Rim countries as suppliers precludes whatever automatic link may have existed between ownership and production location in consumer electronics.

These issues are discussed in chapters 4 and 5. The HDTV activist case is considered first, followed by a skeptic's evaluation.

4
HDTV: Cause for Activism?

In hearing rooms and innumerable position papers, HDTV industrial policy advocates have often justified their call for government support by pointing to foreign HDTV programs and the potential for the consumer HDTV industry to generate interindustry spillovers.[1] The spillover argument has taken two forms. One views HDTV production as a necessary proving ground for a variety of advanced digital technologies and manufacturing techniques (technological spillovers). The other, and more complicated, view is of a U.S. HDTV industry as the critical end-use market for components such as semiconductors and flat panel displays that are widely considered to have pervasive linkages to other electronic segments of the economy (this is an extension of the linkage spillover argument discussed in chapter 2).

But interindustry spillovers alone signal neither an externality problem nor a need for strategic industry promotion. Whether the emerging HDTV industry poses such a question for the nation is examined in two parts. First, this chapter focuses on the spillovers that activists have associated with the development of an HDTV consumer market and the exceptional support that they have called for to develop the industry and to revitalize consumer electronics. Next chapter 5 builds on the framework developed in chapter 2 to question whether the HDTV application of the underlying set of digital imaging technologies merits the strategic industry label on externality grounds and whether the U.S. government has a distinctive role in shaping its development.

What Is HDTV?

Technically HDTV represents the next generation of television, which promises consumers a dramatically improved picture comparable to that of a film and twice as sharp as that of conventional TV sets. It will require much more sophisticated circuitry than the traditional TV set, new production and broadcasting equipment, and a greater allocation of the broadcast spectrum regulated by the Federal Communications Commission. As the next generation of television, the development of an HDTV system is expected to have an impact on every segment of the television industry from program and software production to transmission systems to reception and display equipment. HDTV is now in the prototype stage with full-scale introduction of the HDTV sets and programs not expected in the United States until sometime after the FCC rules in 1993 on the standards that will define American HDTV broadcasts (see figure 4–1).

HDTV also represents a new era in consumer electronics (TVs, VCRs, CDs, and so forth) that will use the more sophisticated electronic technologies and components common to advanced digital computers and information systems. This connection has the potential to create dramatic new opportunities for the consumer electronics and the computer industries to create new products benefiting from each other's innovations, such as HDTV and advanced interactive computers. For TV consumers, HDTV's most immediate promise is a wider set with a sharper image. The longer-term promise is a fully digital, interactive set that will bring the information revolution full force into the living room. The transmission network and receivers for HDTV might become the vehicle for new information services such as video conferences and on-line reference libraries of video and sound recordings. Other applications might allow viewers to tour a house for sale without leaving their own or watch a program on architecture, touch a building on the screen, and access a data base on that structure's history and attributes (table 4–1).

Politically, however, HDTV has come to represent much more. "To miss out on HDTV is to miss out on the 21st century"[2] was a call to action that carried discussion of HDTV from the research labs to the policy front in late 1988, when foreign HDTV development programs were widely publicized in the United States. By early 1990, seven HDTV-related bills had been introduced in Congress, ten committees had held hearings related to HDTV, and countless reports and policy agendas had been circulated by a wide range of interest groups. HDTV also initially received a warm welcome from Secretary of Commerce Robert Mosbacher, who promised that the emerging industry was at the top of his policy agenda and that the administration would institute a national plan for its development. Mosbacher was subsequently reprimanded by White House officials for sounding like an industrial policy advocate, and the HDTV plan was never released. (For a year in the public life of HDTV, see table 4–2.)

At center stage in the policy debate is a constellation of advanced digital technologies that cut across traditional industry boundaries. Interest in these technologies and their commercialization brought

FIGURE 4–1
A High-Definition Television System

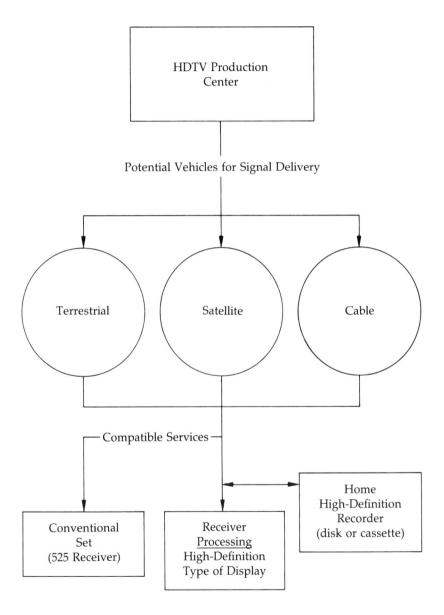

Source: Based on United Nations International Telecommunications Union press release, "International Meeting on HDTV Finds Agreement," April 6, 1990, p. 5.

TABLE 4–1

COMPARISON OF TELEVISION AND OTHER HIGH-RESOLUTION SYSTEMS

NTSC The current forty-year standard used by the United States, Japan, and some others for the transmission of audio and visual information was developed by the National Television Standards Committee. The transmitted signal fits into a single channel or 6 MHz of bandwidth. There are 525 scanning lines, and 4:3 is the width-to-height aspect ratio.

Improved-definition television IDTV receivers, on the market since 1989, requires no change in broadcast equipment or bandwidth, being mainly an improvement in receiver technology. In IDTV the signal received is digitized, processed, stored in memory, and then displayed roughly 60 times per second. These receivers have a relatively small amount of memory. The screen looks like a conventional television screen.

Enhanced-definition television EDTV receivers are expected in the U.S. market within the next five years. The number of scanning lines per television screen may exceed the current standard of 525. Larger screens also may be used. This would be accomplished within the current broadcasting bandwidth of 6 MHz and with much of the current broadcast equipment. The receivers would have more memory and digital signal processing than the IDTV to accommodate the same information.

High-definition television HDTV will use a new signal format based on advanced processing concepts to carry significantly more audio and video information than is possible with the NTSC standard. HDTV receivers will be introduced sometime over the next ten years. They will be similar to EDTV and IDTV, but the displayed image will be twice as sharp and larger. As a result of sending, receiving, and displaying much more information, much more memory and digital signal processing will be required. HDTV will require new production and broadcast equipment as well as a greater allocation of broadcast spectrum. The scanning lines will number between 1,050 and 1,250, with a set aspect ratio of 16:9.

Telecomputer A telecomputer combines television and personal computer functions. The telecomputer will accept a variety of input signals and will permit interactive, multimedia entertainment, education, general data exchange, and communications activities. The on-board microprocessor and memory will allow the telecomputer to perform the computational functions of today's microcomputers. The audio and video quality will be the same as in HDTV, but the flexibility and potential utility will be greater.

SOURCE: Based on U.S. Congressional Budget Office and Defense Intelligence Agency, "High Definition Television Technology" (Preliminary report, November 1989).

TABLE 4–2

A YEAR IN THE PUBLIC LIFE OF HDTV: 1989

February	House Telecommunications and Finance Subcommittee solicits action memos from groups interested in HDTV.
March	House Telecommunications and Finance Subcommittee holds hearing on HDTV in which Commerce Secretary Robert Mosbacher said that the "highest priority" would be given to the creation of a national HDTV plan. (Mosbacher was later chastised for sounding like an industrial policy advocate in opposition to the official policy of the Bush administration.)
	House Committee on Science, Space, and Technology holds hearing on HDTV.
May	Senate Committee on Commerce, Science, and Transportation holds hearing on two HDTV bills.
	American Electronics Association releases its industry-led proposal for a $1.35 billion commitment from the government to support HDTV and the ATV Corporation.
	House Armed Services Subcommittee on R&D holds hearing on HDTV.
	Senate Subcommittee on Science, Technology, and Space holds hearing related to HDTV.
	West German telecommunications minister proposes to Commerce Secretary Mosbacher U.S.-EC cooperation on development of high-definition television to catch up to Japanese HDTV producers.
	Zenith's chairman proposes a consumer tax on color televisions to raise funds for HDTV research.
	"Wake Up America!" conference is sponsored by American Electronics Association, National Center for Manufacturing Sciences, and Rebuild America to promote the industry-led targeting strategy and the AEA's strategy in particular for commercializing HDTV through the public-private ATV Corporation.
	House Subcommittee on International Cooperation holds hearing on HDTV and international standards.
June	Senate Committee on Commerce, Science, and Transportation holds hearing related to HDTV.

TABLE 4–2 *(Continued)*

	DARPA announces first selections—five companies—for its High-Definition Display Program.
July	Senate Committee on Commerce, Science, and Transportation holds hearing related to HDTV.
	House Subcommittee on Economic and Commercial Law holds hearing related to HDTV joint production ventures.
	Congressional Budget Office reports importance of consumer HDTV industry has been overstated.
August	Senate Committee on Governmental Affairs holds hearing on prospects for developing a U.S. HDTV industry.
September	House Subcommittee on Telecommunications and Finance holds hearing on HDTV.
	Release of the Commerce Department's HDTV plan delayed indefinitely, and "HDTV Working Group" changed to "Technologies Working Group" to focus on the "range of technologies" rather than the HDTV industry alone.
October	DARPA announces selection of three more contract teams for High-Definition Display Program.
November	The President's National Advisory Committee on Semiconductors calls for revitalization of the consumer electronics industry to boost demand for U.S. semiconductor sales through the creation of the Consumer Electronics Capital Equipment Corporation.

SOURCE: Author.

attention to the HDTV application, but the claims and proposals of activists quickly altered the focus and made the creation of a U.S. HDTV industry and a consumer electronics industry the policy issue. To investigate the various claims and proposals for strategic industry promotion, a clear policy focus is needed: what is the product and industry that the government is being asked to label as strategic and promote?

Cross-cutting Technologies and the HDTV Industry

No one doubts that advanced imaging technologies or the technologies underlying high definition systems (HDTV is one type) will have a

TABLE 4–3

SELECTED APPLICATIONS OF HIGH-RESOLUTION SYSTEMS

Field	Application
Medicine	
Pathology	Long-distance diagnoses
Medical education	Real-time observation of surgery
Space	
Launch control and evaluation	Rapid and improved technical analysis
Space station freedom	Precise execution of tasks in unfamiliar surroundings
Humanities	Easy public access to array of art works stored on laser disks
Printing and electronic publishing	Reproduction of TV images as magazine-quality pictures
Telecommunications	Improved data transmission for teleconferencing
Industry	Improved images for CAD and CAM systems
Computers	Higher quality in computer-generated graphics

SOURCE: U.S. General Accounting Office, *High-Definition Television: Applications for This New Technology* (Washington, D.C.: Government Printing Office, December 1989).

wide range of applications in entertainment, business, and research.[3] The U.S. General Accounting Office in a December 1989 report cited fourteen different existing and emerging applications of high-resolution imaging systems. In the field of pathology, for instance, demand already exists for clear, high resolution images. (See table 4–3 for other examples.)

There is no reason to presume, however, that a domestic consumer electronic industry led by HDTV receiver production will drive the development of these technologies, other commercial applications (end-use markets), and ultimately national economic competitiveness. Nor is there any reason to presume that the investment and strategies of participants in the television industry are inadequate from the perspective of national welfare so that the government needs to intervene with plans for special assistance. Yet industrial policy proposals have focused on the consumer HDTV industry as the critical end-use market and the launching pad for the development of the set of cross-cutting technologies.[4] This approach begs the question of the net benefit of this focus relative to other uses of public and private resources. At issue are whether the United States needs to produce a particular type of hardware and to revitalize the domestic consumer electronics industry and whether strategic industry promotion is the most effective way to ensure this outcome: does the United States need to speed and to encourage actively through specialized incentives a domestic HDTV set industry?

This issue is repeatedly confused in the policy debate. Part of the problem is the mix of HDTV definitions that have become entangled in policy discussions: a consumer electronics product, a video service for the next information age, a high-resolution imaging technology with multiple applications, as well as the critical dynamic link in the electronics food chain.

Since strategic industry prescriptions of HDTV activists have focused on the need to create special incentives to create a consumer electronics industry led by HDTV, this study also focuses on HDTV as a consumer electronics product to examine the merits of their claims and proposals.[5]

How support should be structured for advanced technologies also raises the larger issue of false dichotomies, which seem to be characteristic of arguments favoring domestic production in a particular industry. The HDTV case for strategic industry promotion, for example, has often been described in terms of either targeting consumer HDTV now or forgetting about the underlying set of technologies. Supporting analysis then has compared the benefits of developing these technologies through a promotion plan for the HDTV industry with a null set in which the technologies are not developed or not available in the United States. Chapter 5 demonstrates that this is a serious misrepresentation of the intricate and multidimensional nature of the development process.

Given that the attractiveness and potential of the underlying technologies are a sound indicator that even without government intervention the technologies will be developed, the question is whether the United States can expect to share in the technological and commercial returns from the development of advanced technologies. How policy can help is the subject of the discussion in chapter 6.

To understand the forces that brought HDTV to the policy front in late 1988, some background on the HDTV programs abroad is needed.

HDTV in Japan and Western Europe

In search of new consumer electronic products, the Japanese public broadcasting company NHK noticed some weaknesses in conventional television and decided in the early 1970s to try to discover an avenue for increasing viewer enjoyment. The Japanese found a positive response to increasing the viewing angle and picture resolution. To take advantage of this potential market, NHK has spent more than $700 million over the past twenty years to develop HDTV.

Their resulting Hi-Vision system extends conventional TV analog technology and is based on the signal compression technique known as MUSE, which was officially adopted in 1984 by Japanese transmission equipment manufacturers.[6] HDTV programming is, however, limited and can be viewed only for an hour a day in Japan via direct broadcast satellites. Full-scale broadcasts have been postponed until the summer of 1991. Developing attractive software for Hi-Vision appears to be the major technical stumbling block in the development of the Japanese HDTV system.[7] To view HDTV broadcasts, Japanese consumers will also need to purchase new sets that were recently introduced at $34,000 by Sony, Matsushita, and Hitachi.[8]

The array of actors involved has created several dimensions to the development of HDTV in Japan. At one level is the bureaucratic rivalry between the Ministry of Trade and Industry and the Ministry of Posts and Telecommunications. The two are competing for influence in the research and diffusion of the technologies underlying HDTV. They differ in their views on the extent and pace of HDTV household penetration. MITI and equipment manufacturing firms believe that nonconsumer applications of high-definition systems will precede consumer HDTV markets. The applications in areas such as medicine, printing, and the arts are expected as the launch markets for descending the learning curves in the underlying technologies and components.[9] On the flip side MPT does not consider the high cost of Hi-Vision a major obstacle to consumer demand, but it is troubled by the prospect of limited programming over the next four years.

The rivalry between the two ministries is being played out through their R&D and promotion programs. Each reportedly plans to invest $115 million in public relations and user promotion programs. Each is also funding the development of display technology. MITI has focused on large, flat panel displays through the Giant Technology Corporation; MPT has focused on projection displays through the Advanced Image Technology Research Center.

As the central player in Hi-Vision, NHK and its research labs are also in the competition for influence, challenging the role of MITI and

MPT in the development of HDTV and the general promotion of high technology in Japan. To avoid being overshadowed by NHK, private Japanese stations have also entered the competition with the development of EDTV or Clear-vision as an intermediate step to HDTV.

Nippon Telegraph and Telephone Corporation also has an interest in the development of consumer HDTV as a path to broaden its market base. NTT's interest is the delivery of video signals to homes and businesses to offset the cost of building a broad-band network. The business market is expected to be larger because of the greater demand for advanced services such as digital data transmission and voice telephony. Serving household entertainment needs expands the use of networks and thus could help lower the cost of building the network as well as the cost of providing business services such as video for teleconferencing.[10]

A similar interest in scale opportunities led the Japanese to propose MUSE Vision as the international standard. But the proposed adoption of the Japanese production standard system was firmly rejected by the Europeans at the 1986 Consultative Committee on International Radio (CCIR) meeting, the international group responsible for production standards, and prompted the formation of a European research group led by Philips, Thomson, Bosch, and Nokia of Finland to develop a European HDTV system. The development of this alternative standard reportedly cost $350 million.[11] At least for the foreseeable future, this effort has locked the Europeans into a standard referred to as the multiplexed analog components (MAC) system.[12]

The European HDTV project, as part of the European Eureka program, is the public-private consortium that was held in U.S. congressional HDTV hearings to be a model for U.S. industry-led strategic targeting programs. In July 1990 the HDTV project entered the second phase of a ten-year plan, with $600 million committed to full commercial production of a range of high-definition consumer electronics equipment (television sets, domestic VCRs, video disks) by 1995. Technologies of particular interest are non-CRT displays such as liquid crystal displays and digital recording techniques. Outside the Eureka framework Thomson and Philips signed a general agreement in June 1990 to explore HDTV options, together committing $3.8 billion over the next five years.

Similar to their American counterparts, European producers face an uncertain consumer market. According to an electronics industry analyst at BIS Mackintosh in the United Kingdom, "by the end of a decade there will be a sizable percentage of European homes with a high-definition receiver. The unanswerable question, though, is what year we shall start the decade."[13]

47

The Japanese and Europeans have been trying to start the decade but have encountered problems with their choice of standards (see the Appendix). Neither the Japanese nor the European system is designed to be compatible with the current U.S. broadcasting environment, which relies on over-the-air transmissions of television signals. As a result, the FCC ruled in 1988 that the HDTV transmission system selected for the United States must be compatible with the existing stock of 160 million receivers of the National Television Standards Committee (NTSC) system. Further, the European and Japanese systems are based on updated versions of conventional analog technologies, while digital is widely expected to be the wave of the future in video technologies and products.[14]

HDTV as a Strategic Industry Candidate

Inspired by these developments abroad, HDTV activists in the United States have argued that "America cannot afford to sit on the sidelines and let other nations develop an HDTV industry without our participation."[15] Activists have heralded HDTV as the competitiveness barometer for the entire electronics industry and the nation so that emerging a winner in the HDTV industry would demonstrate the economic health of the United States. First pointing to government-supported HDTV programs in Japan and Europe, they held that the United States was already behind. Then pointing to Zenith as the only major U.S.-owned television manufacturer, they contended that the United States was perilously close to falling even further behind. They concluded that without special help from the government to reallocate resources and to speed the creation of a domestic HDTV industry, HDTV would be the first in a series of industrial dominoes to fall: foreign HDTV promotion programs would make it impossible for the United States to capture the returns from commercialization of the underlying technologies, the market shares of computer and electronics equipment manufacturers would suffer as a result, and U.S. technological and economic progress could ultimately be curtailed.[16]

The activist argument is the following: the consumer market for HDTV products (receivers and VCRs) will be a large market for sales of semiconductor and other advanced components, supplying the HDTV market will generate productivity improvements in the upstream industries and in the national economy, U.S. suppliers will be able to participate only if the HDTV is both owned and controlled by U.S. firms, and unless U.S. firms participate in HDTV, the competitiveness of other products such as computers that use the technologies and components in HDTV products will decline as will the technological competitiveness of the nation. The case concludes that the creation of a large-scale HDTV set industry is therefore necessary to revitalize U.S. consumer electronics, to breathe new life in the U.S. semiconductor industry, and to promote national economic health.[17]

The special importance of HDTV is connected specifically to the rapid response time and immense video information-processing requirements of HDTVs and the prospect of a large-scale household consumer market that will demand low-cost products. The first separates HDTV from past generations of consumer electronics equipment. Together with the prospective consumer market, activists believe the emerging HDTV industry has the potential to drive advances in the underlying manufacturing, semiconductor, and display technologies and through them the computer industry. They further suggest that strategic industry promotion of HDTV should be viewed as promotion not of a single industry but of an essential link in the electronics food chain that will help a wide range of U.S. firms capture productivity-enhancing benefits that would not otherwise be captured from the development of advanced digital technologies and components.

Technological Spillovers. One advantage that has been associated with HDTV production is the knowledge that will be generated on flexible manufacturing techniques. Consumer electronics has been characterized by high volume production, low profit margins, and short-product cycles (eighteen to twenty-four months), which together have placed a premium on efficient, flexible production that can respond quickly to changing market conditions.[18] Some believe that competitive strength in consumer electronics generates critical generic manufacturing skills and technologies for the national economy and that the decision of U.S.-owned firms to exit the industry narrowed the manufacturing skill base of the U.S. economy. From this perspective HDTV is viewed by activists as a rare market opportunity to stimulate consumer electronics production in the United States and thereby to generate improvements in the pool of manufacturing knowledge.[19] Proponents also expect HDTV to generate significant technological spillovers by pushing the frontiers of advanced semiconductors and video display technologies, which will then feed into other downstream electronics user industries.

Linkage Spillovers. For an industry that accounts for less than 0.1 percent of U.S. manufacturing, consumer electronics seems an odd cornerstone upon which to build national competitiveness (table 4–4).

49

TABLE 4–4

SELECTED U.S. MANUFACTURING INDUSTRIES,
WITH VALUE ADDED AND EMPLOYMENT
(percent)

Industry	Value Added	Employment
Automobiles	5.7	4.3
Chemicals	9.7	4.7
Commercial aircraft	3.6	3.3
Computers, semiconductors, and office equipment	3.7	3.1
Consumer electronics	0.3	0.7
Machine tools	0.3	0.3
Steel	1.5	1.5
Textiles and apparel	3.4	4.0

SOURCE: U.S. Bureau of the Census, *Annual Survey of Manufactures, 1986* (Washington, D.C.: Government Printing Office, 1988) as reported by MIT Commission on Industrial Productivity in *Made in America* (Cambridge: MIT Press, 1989), p. 8.

Some counter that this perspective underestimates the real value of the consumer electronics link in the electronics food chain. Representative Ritter, in pressing the candidacy of HDTV, has argued that "if the US does not compete in consumer electronics, it will not have a semiconductor industry. Without a semiconductor industry, we will not have a state of the art computer industry or a military electronics industry."[20] The argument assumes that a thriving computer industry cannot be based on foreign technology or components and that a U.S. HDTV industry will secure a domestic base in components by creating a large market for sales from U.S.-owned companies.

The HDTV interindustry spillover argument is an extension of the Swiss watch case discussed in chapter 2. In that case the machinery used for watch production is characterized by increasing returns to scale (average costs decline as output is increased). Consequently, as watch production expands and machinery purchases increase, the unit price of machinery falls, and society enjoys a real increase in resource savings.

The strategic industry case for HDTV argues that it is the most important end-use market for the upstream semiconductor and flat panel display industries, which are characterized by increasing returns to scale frequently associated with dynamic learning effects (a decline in average cost as cumulative output increases). Consequently, as demand and production of HDTV sets increases, component purchases will increase, and the unit price of chips, displays and sets will fall. Any input cost-savings of inputs from HDTV production may furthermore spread to a variety of other downstream products outside the entertainment business, such as computers and telecommunication.

Upstream industries. The first link, and the most prominent one, in activist arguments, is the role of consumer HDTV for the upstream semiconductor and display industries. The demand for video image-processing capability will, for example, require HDTV receivers to have more memory and more complicated circuits than currently required in conventional NTSC sets.[21] HDTVs also will require the rapid processing of large amounts of video data to handle a real-time HDTV full-color signal that can contain up to 1.2 billion bits per second.[22] Fitting this HDTV signal into the existing allocation of bandwidths[23] in the United States will thus rely on the compression of data before transmission and decompression upon reception, or in the words of one engineer "squeezing an elephant into a bathtub." This task is accomplished by specialized chips and digital signal processors. Some analysts believe that if HDTV becomes a popular consumer electronics product, it will provide the scale necessary to drive the learning economies associated with sophisticated semiconductor development and production, generating cost savings not only for HDTV but also for the other downstream industries.

But HDTV presents a bit of a chicken-egg problem in that many experts do not expect HDTV to be affordable for most consumers until large flat panel displays become affordable. Firms in the United States, Japan, and Europe are in a race to develop extra-sharp flat panel displays.[24] For high information content displays, there are five competing technologies: active liquid crystal displays, plasma technology, electroluminescence, vacuum microelectronics, and deformable mirrors.[25] Of these, active matrix liquid-crystal displays have received the most attention. The president of Stanford Resources Inc. (SRI), Joseph A. Castellano, has characterized it as a large integrated circuit deposited on glass. The larger the display, the more active elements are required, and the greater the chance for defects.[26]

As with conventional television, the display component for HDTV is expected to account for 30 to 50 percent of a receiver's production costs. The large size and weight of HDTV sets based on cathode ray tubes will tend to keep retail prices high and out of reach of the average consumer. Technological and manufacturing advances in flat panel displays will consequently be essential for minimizing the weight of the receivers and the consumer cost of advanced television. One of the main technological hurdles in the path of commercial

HDTV development is thus to increase production yields of large displays while reducing prices to affordable levels.[27] This will require precision manufacturing analogous to that required for sophisticated semiconductor production. The hope is that the consumer HDTV market will provide the scale to launch display suppliers over the technical hurdles and subsequently to generate benefits for other downstream industries.

HDTV and the Digital Revolution. The ongoing revolution in digital technologies and applications sweeping the electronics industry serves as the technical framework that links HDTV and other electronic end-use markets. This revolution is providing a common digital language (a series of zeros and ones) for the transmission, processing, and display of all types of information. Technical experts expect that the eventual emergence of a single route for data transmission will lead to widespread use of information technologies and possible convergence of applications. In the computer industry, for instance, telecomputers are expected to integrate sound, video, and data sources and consequently to be capable of handling a variety of entertainment, education, and business functions. Paralleling this, the HDTV path extends from the consumer electronics side of the electronics tree. It is also expected to lead ultimately to an interactive digital set that allows viewers to modify and interact with broadcast programs.

As television shifts toward digital electronics, the opportunities for technological linkages with computers is expected to increase. The development of a fully digital HDTV system—end-to-end digital transmission—is central to this link and the development of interactive television. Once in digital form, information can be stored and manipulated by a computer. The common (digital) language promises to expand the two-way traffic in spillovers between consumer electronics and computer products. The dual-use nature of many of the technologies and components that make up television has historically created the opportunity for a close sharing of technologies with the computer industry. But the opportunities are expected to expand further as computers move toward digital video computing and TV sets adopt computer functions. In particular, engineers expect that advanced TVs will benefit from the digital integrated circuit and semiconductor memory technology developed for the computer industry, while advanced computers will benefit from some of the digital video display technology and high-resolution screens developed for the television industry.[28]

Some have taken convergence in technologies to mean market convergence. Instead of multiple terminals, household consumers would perform multiple tasks with a single terminal. In this context Japanese HDTV has been portrayed as a dangerous competitor for the rapidly emerging array of multimedia personal computer products: HDTV (them) versus industrial PCTV (us). The concern is that Japanese HDTV producers will have the benefit of higher production volumes for the consumer HDTV product and undercut the international cost-competitiveness of advanced American computer products.

Another, and perhaps most important, link concerns the connection between telecommunications and HDTV. Broadly defined, telecommunications is both a service activity and a manufacturing industry. As used in this study, telecommunications involves point-to-point communications (voice, data, and video), which excludes broadcasting and includes infrastructure and network equipment (switching, transmission, and terminal equipment), facilities services (renting of leased lines or circuits by users), and transmission services.[29]

Consumer HDTV has been associated with advances in telecommunications equipment because of the advances in digital signal processing and complex analog/digital converters needed for HDTVs. The high data-processing rate planned for HDTV exceeds the capacity of existing telephone lines. Advances in transmission technologies to circumvent this problem are expected to improve the data transmission pathways for traditional computer and communications users. Further, networking design problems that would need to be resolved for the use of HDTV receivers as interactive (two-way) terminals may reinforce or prompt advances in communications technology. From this perspective, prospective consumer HDTV demand represents an opportunity to expand demand for broadband telecommunications networks, particularly to the extent that household consumers demand interactive terminals to serve both their entertainment and computer needs.

Some observers have thus argued that the HDTV consumer industry will function as a pump primer that will speed the unraveling of network design problems and the creation of a public fiber-optic highway for the information industries of the future.[30] Others remain skeptical, uncertain about the extent of demand for interactive versus passive (one-way) television and the importance of HDTV's role of pump primer relative to other national interests such as creation of the underlying public data highway itself.[31]

Jeffrey Hart and Laura Tyson aptly summarize the various HDTV spillover arguments discussed above: "The greater is U.S. participation in HDTV consumer markets, therefore, the greater will be the upstream, downstream, and manufacturing benefits for the rest of the

U.S. economy. Thus, policy measures should be aimed at maximizing U.S. participation."[32] Among other problems, policy makers must contend with a wide array of interest groups and action agendas that seek to lead them on how this participation should be encouraged.

Military Security and HDTV

Before turning to the various HDTV proposals, a brief detour is in order to make a clear distinction between the objectives and strategic industry arguments based on economic criteria from those concerning military security interests. National security interests have been linked to high-definition systems because of the expected heavy reliance of the military on high-definition display technologies and components in areas such as command and control, training and simulation, and intelligence analysis.[33] To the extent that inputs such as flat panel displays are considered critical for military objectives such as sustaining technological superiority, Defense may conclude that the industry is strategic.

If, however, the objective of strategic industry promotion is to enhance economic welfare, then the criteria for the military's strategic industry label should not be substituted for the merits of the case for strategic industry promotion on economic grounds. The military should not be used as a backdoor to gain national support for an industry promotion plan that failed to marshall the necessary economic arguments and evidence (that is, reasonable evidence of inadequate investment in the industry in question to generate the interindustry spillovers that are expected to play a key role in the economic future of the United States) to support the case for special attention.[34]

This study examines *only* the economic case for strategic industry promotion of HDTV. It leaves military concerns to other analysts. The merits of the consumer HDTV argument will depend on factors such as the relative importance of domestic HDTV manufacturing, the actions of private actors to take advantage of the spillovers that experts believe with reasonable certainty will become important, and the distinctive role of government in supporting the development of the advanced technologies and in encouraging the diffusion of spillovers.

Proposals and Empirical Support

Given the special importance of HDTV, aggressive targeting programs abroad, and only one U.S.-owned television manufacturer in the United States, HDTV activists concluded that the government commitment to HDTV would have to be exceptional. The government would need to assure American firms that invest in HDTV that they will get both the government's HDTV business and a partnership for developing the domestic industry. At the prospect of a federal partnership for action, HDTV inspired about as many calls for help as definitions that vary in intent, emphasis, coverage, and tools. Specific suggestions included grants, guaranteed loans, special taxes, infant-industry nurturing and protection, as well as procurement contracts to guarantee domestic producers an initial market.[35]

The early stage of the debate focused on HDTV as an innovative new technology embodied in advanced TV systems and receivers and the need for some form of public-private HDTV equipment manufacturing consortium to revitalize the domestic consumer electronics industry.

Proponents have described the goal of such a HDTV industry-led strategy as the development of "a U.S. owned and controlled HDTV industry that ensures that such HDTV components as chips and displays, and such related HDTV products as VCRs and computer monitors, are manufactured here by U.S.-owned companies."[36] In other words the intent is similar to the European one: an independent domestic capability to produce HDTVs. Proponents warn about dependence on foreign firms that benefit from closed home markets, for advanced components and technologies. They propose that domestic production of a final product such as HDTV will create a greater sense of economic security by improving the sales outlook for producers of upstream components.

This position is perhaps best reflected in the sweeping $1.35 billion proposal advanced by the American Electronics Association in the spring of 1989.[37] The association had proposed earlier in the year that advanced television receive "national champion"-style treatment.[38] A federally chartered development corporation, the Advanced T.V. Corporation, was the cornerstone of the subsequent industry-led HDTV promotion plan. The corporation's objective was to channel public and private funds away from activities with presumably lower economic value toward the creation and nurturing of a domestic HDTV industry. This was to be encouraged by stimulating demand, conditioning access to loans from the ATV Corporation on local semiconductor content as well other performance requirements, and basing U.S. market access on reciprocal access abroad (table 4–5).

The less than enthusiastic reception of this proposal was followed by a more ambiguous one advanced by the President's National Advisory Committee on Semiconductors. The stated objective was once again the creation of downstream markets to expand the market share of U.S. semiconductor manufacturers.[39] Although not mentioned

TABLE 4–5

$1.3 BILLION AEA PLAN FOR HDTV

Action	Federal Cost	Purpose
Grants	$300 million over three years	To develop basic technologies, including advanced displays and image-processing chips for an advanced TV infrastructure
Loans	$500 million over 5–10 years	To discount loans to offset the cost of capital, for investment in manufacturing and product-development strategies
Loan grantees	$500 million over 5–10 years	To encourage financial institutions to invest in HDTV ventures
ATV Corporation		To manage funds based on semiconductor local content requirements as well as R&D and production performance requirements, to stimulate consumer demand, and to oversee standards, licensing, and import activities
Conditional access to U.S.		To base access to U.S. on reciprocal access in other markets

SOURCE: American Electronics Association, "Development of a U.S.-Based ATV Industry," May 9, 1989.

directly, HDTV was at the heart of the proposed Consumer Electronics Capital Corporation (CECC) intended "to resurrect the U.S. consumer electronics industry infrastructure by providing a multi-billion dollar pool of very patient, low-cost capital." The role of the government would be "purposefully [to] support market re-entry for U.S. consumer electronics and improve its potential to compete vigorously," to permit access to the U.S. consumer electronics market only in return for access abroad, and to set guidelines for both local technology content and R&D performance requirements. Federal financial support was vaguely defined as "pledges of support," but the clear intention was to shift capital and labor resources toward the domestic production of consumer electronics products such as HDTV.[40]

A mixture of other recommendations has been advanced to promote the development of HDTV as a consumer electronics product. These have included user fees on conventional television to subsidize development of advanced television as well as import restrictions and selective procurement practices to nurture the development of a domestic industry. In the 1989–1990 period five bills—none enacted—singled out HDTV and called for the promotion of research, design, development, manufacture, and commercialization of HDTV in the United States.[41] The 1990 competitiveness package proposed by the Democrats included financial assistance for HDTV development as part of $400 million spread over three years for the Advanced Technology Program in the Commerce Department. ATP is intended to help companies commercialize discoveries and refine manufacturing techniques. Other elements included $80 million per year for the development of dual-use technologies through defense-related projects and the creation of a government-industry board to oversee the development of high-resolution systems in the United States.[42]

As discussed, some refer to HDTV as a set of advanced digital technologies with pervasive applications. The federal role here is usually discussed in terms of research support for "generic" or "enabling" technologies. As with the term "strategic," few well-defined policy rules and procedures exist to evaluate objectively and systematically what support for generic and enabling technologies should entail in specific project requests. In the call for domestic HDTV production, frequent references to HDTV as a generic technology fundamentally confuse the industrial policy debate by lumping together national interests in the set of technologies that cut across industry and national boundaries with a specific application, namely, a domestic HDTV industry.

Evidence on Expected Consumer Market and Spillovers. The early quest for data to evaluate the future market importance of consumer HDTV centered on three 1988 reports commissioned by the Commerce Department, the Electronics Industry Association, and the American Electronics Association.[43] Reliable research data about the consumer market for advanced television systems were scarce. As a result each forecast was based on analogies to past successful consumer electronics products. The forecasts used different sets of assumptions on price, rates of consumer acceptance and household penetration, and the extent to which intermediate products such as improved-definition TV (IDTV) and enhanced-definition TV (EDTV) diffuse demand or take the wind out of the sales of HDTV. The estimates for the HDTV receiver market in the year 2003 subsequently range anywhere from AEA's $5 billion to EIA's $20 billion (see table 4–5). Once HDTV VCRs are included, the estimates range from the AEA's $8 billion to $15 billion based on the Commerce Department and Darby's high-growth scenario or $11 billion if the mean of the two (high- and low-growth) forecasts is used.

A common belief in the consumer success of HDTV links the three forecasts. All are relatively optimistic about the rate of household penetration. With the exception of the low-growth scenario in the Darby report (which receives relatively little attention in the report), the studies assume that the adoption rate of HDTV will approximate that of color TVs or VCRs. But if consumers perceive less of a difference between conventional (NTSC) television and HDTV than in the transition to color, then the adoption rate of HDTV may be slower. Similarly the relatively rapid adoption of VCRs (which complemented rather than substituted for TVs) may not extend to HDTVs if consumers on average do not perceive new service value from owning an HDTV.[44] (Chapter 5 discusses in more detail consumer demand factors.)

Few empirical estimates of potential spillovers are available. Based on its calculations the AEA figures that U.S.-owned firms will need 50 percent of the U.S. HDTV market just to maintain current worldwide market shares in computers and semiconductors. If the U.S. HDTV market share is less than 10 percent, U.S. shares of the global computer and semiconductor market will be cut in half to roughly 35 and 21 percent, respectively.[45] In concert with these claims, AEA's rallying cry at congressional hearings has been that "we must gain an industry to retain an industry."[46] By AEA calculations this requires tripling Zenith's market position for color TV of 12 percent—an impossible task.

Moreover the method in which the statistics were generated leaves the link between HDTV and the success of other industries at best uncertain. When pressed to explain the AEA methodology, Richard Iverson, AEA president, told an HDTV conference that the projections were not based on an underlying model and they were "not done very analytically."[47] Instead the statistics represented the informed judgments of a panel of experts assembled by the association. Yet the pace and breadth of the digital revolution continue to surprise and confound even the experts.[48]

Predictions of the economic importance of an emerging end-use market such as HDTV in this revolution are thus inherently imprecise and uncertain. The AEA predictions should therefore be acknowledged as tenuous and the assumptions (on consumer demand, relative industry growth rates, ownership, production location, and the role of strategic alliances) made explicit. Furthermore, their use should be restricted to broad conjectures of potential industry trends rather than to support sweeping statements on the extent of future commercial interdependence and the need for stratgic industry promotion. To their credit some HDTV supporters have noted that the statistics and the magnitude of spillover effects from an HDTV consumer industry are uncertain.

The offered fallback position, however, does little to advance the strategic industry case or the industrial policy conclusion: if others are promoting HDTV, it must be worthy of promotion by the United States. As discussed in chapter 2, the logic of this position is elusive. Even if the relative economic importance of spillovers from a consumer HDTV industry were known, the federal government, by coaxing U.S. firms to enter and simply to follow the leader into a competitive and crowded marketplace, is unlikely to accomplish much beyond the creation of excess capacity in the global industry.

Chapters 5 and 6 also show that it is not possible to conclude with reasonable certainty that consumer HDTV is any more important for American economic welfare than computers or telecommunications or that private participants are failing to take advantage of potential linkages between the various industries. It is therefore unclear why the nation would gain any additional benefit from reallocating resources to develop "indigenous HDTV technologies"[49] and to speed the development of a domestic HDTV industry.

5
Cause for Skepticism

No one is seriously questioning whether a number of the technologies and components that underlie the HDTV industry will have applications in an array of other industries: clearly they will. The starting point for evaluating the merits of targeting HDTV for promotion as a strategic industry, however, must be the reverse; namely, what is the nature of the linkage between consumer HDTV and other applications of the underlying technologies? Recall that the key assumptions in the activist case are the following: an American HDTV industry that made TV sets would be the essential final-product market for both cultivating technologies with multiple applications and stimulating sales for the U.S. semiconductor industry; HDTV market shares would determine the commercial competitiveness of other electronics industries and ultimately of economic progress in the nation; and most important, without special government assistance to match programs abroad, the American HDTV industry would not develop, or at a minimum be delayed, and the nation would subsequently miss out on the returns from developing the underlying technologies. Or as Rep. Don Ritter (D, Pa.) concluded, "for America to miss out on HDTV is to miss out on the 21st century. It's as simple as that."[1]

But is it really that simple? Many have their doubts. The Advisory Committee on Advanced Television Services of the Federal Communications Commission questioned whether the American Electronics Association was correct to assume that future success in computers depends on a U.S.-owned and -controlled HDTV and VCR industry. After reviewing AEA claims in 1989, the committee concluded that "at this juncture, it is impossible to forecast the technological spillovers from U.S. HDTV development. Nor is it clear why U.S. manufacture of PCs and automated manufacturing equipment cannot thrive even if the U.S. ATV system is based on a foreign technology."[2]

Other legitimate concerns about the industry-led case for promotion of HDTV exist. Should the United States speed or favor the development of the HDTV industry if consumer demand is uncertain? How can the U.S. strategy to create a domestic consumer HDTV industry be led by industry if participants in the U.S. television industry cannot agree on a common agenda and if major U.S.-owned electronics companies are more interested in developing the computer applications of advanced digital technologies? Why should the government risk undermining the emerging competition between the various high-definition systems (such as HDTVs engineering workstations, multimedia PCs) being developed by targeting one for preferential access to credit? Why are the current actions and investments of those developing the domestic HDTV industry inadequate? Given the uncertain need to promote domestic HDTV production on technical grounds, HDTV strategic policy initiatives seem particularly well designed to stimulate an upward spiral of "I'm strategic too" lobbying in the United States and a matching spiral of rising subsidies at the international level.

These issues are relevant because they help to put the claims of HDTV activists into perspective. The policy debate over HDTV is not about whether government promotion would be sufficient to advance U.S. economic interests: activists are quick to concede that macroeconomic and other microeconomic adjustments will be necessary. At issue instead are whether industry-specific promotion plans are necessary and whether they represent the best use of private and public resources.

Industry-led initiatives clearly start the process of inquiry, but they cannot resolve the difficult technical, economic, and political issues that need to be addressed if intervention is to be welfare enhancing for the nation. To address the issue of strategic spillovers and the necessary policy response honestly, a rigorous analysis of the interdependent relationships between the various technologies, components, industries, *and* investment strategies would need to be undertaken. To determine the nature of the interindustry relationships and their possible evolution over time, such an analysis would first need to determine the direction and importance of technological exchanges between the industries while taking into account outside factors that influence the relationship but are not determined by an individual industry. Given these findings, the analysis would then need to determine the extent of underinvestment in the critical spillovers generated by the industry or application that is being considered for special treatment. Next the analysis would need to determine the cause of underinvestment and whether it could be addressed through federal subsidies and if so, additional questions include how much and how long subsidies should be given.

With the HDTV industry the discussion has not progressed beyond the first stage because of disagreements among technology experts. More important, the other steps critical for building a base for sound economic prescriptions have been largely ignored in the vast volume of paper produced on HDTV and the need for strategic industry promotion. This chapter accordingly finds more reason to suspect than to support the industrial policy conclusion of activists. First, the strategic importance of consumer HDTV has been overstated by activists; the debate among technology experts over the commercial dependence of other industries on consumer HDTV continues today. Second, the threat from foreign HDTV programs has been exaggerated.

This chapter explores these concerns by first examining the nature and expected degree of consumer demand. Next the discussion turns to the role of HDTV relative to the role of industries presently outside the entertainment business in developing the core digital technologies. A third section broadens the discussion to consider the role of international actors and activities in generating the spillovers that have been associated with HDTV and other high-definition systems. The last two sections take a step back to examine the practical difficulties of designing in the 1990s an industry-specific approach to promote national technological capabilities.

How Certain Is Consumer Demand?

The importance of HDTV as an end-use market is contingent on the size and growth rate of this market relative to other applications of advanced technologies and components. Consumers must be attracted to HDTV in sufficient numbers to generate an all-important mass market that can generate the learning and spillover benefits for other downstream industries. But the timing and perhaps the ultimate importance of this market for other industries are questionable.

The problem is partly one of expectations. The last consumer market success story was the VCR, which was original in nature and complementary to existing TV sets. The market test for HDTV will be whether it can sufficiently impress consumers with dramatic improvements in service value so that a mass market develops. Many questions exist, however, about the nature of potential consumer demand. First, the market potential of emerging industries is inherently difficult to forecast because both demand and price are unknown, simultaneously determined variables. This problem is particularly acute for HDTV because the system will require both dramatic advances in displays and technologies and changes in the way consumers watch television. Several HDTV reports have as a result concluded that consumer demand for HDTV is uncertain. Unresolved issues include the extent to which the benefits of HDTV will be apparent to consumers and the extent to which they will on average pay a premium for the higher resolution, wider screen, and digital stereo sound of an HDTV set: initially $2,000 for a HDTV to replace their $400 conventional color television. Prices will fall as production increases, but whether and when they will fall enough to induce large numbers of consumers to switch to HDTV is a matter of contention.

The evidence on the possible timing and degree of consumer demand and price reductions is mixed.[3] For the transition from black-and-white to color sets, consumers were willing to pay the higher prices for the new sets, and the prices declined as household penetration increased. The transition to HDTV would differ in two important respects. First, available intermediate products such as EDTV may give consumers the option to trade off improvements in image resolution against cost and therefore may reduce total potential demand for HDTV. If a significant share of consumers settle for a less sophisticated set, then the scale of the consumer HDTV market would be reduced (at least for the near term), and the manufacturing and other spillover benefits might also be reduced.

Second, preliminary market research at MIT suggests that HDTV sets will need to be considerably larger than conventional sets for the benefits of HDTV to be apparent to the average viewer.[4] This research also suggests that viewing conditions and program content will affect consumer enthusiasm for HDTV. In particular the MIT study indicates that despite years of market research and twenty years invested in HDTV at a cost of more than a billion dollars, U.S. consumers are still not interested in the HDTV system developed by Japan. Experts speculate that consumer demand for HDTV will take off and prices will fall dramatically once large bright flat screens are widely available. But these screens are little more than a blip on the horizon. Furthermore, the degree to which consumers will prefer much larger receivers (40 inch and up) to sit in the living room or hang on the wall once the flat panel technology is viable, is undetermined.

TV programs in the HDTV format will also need to be widely available before a mass consumer audience is likely to develop. But producers may wait for the market to develop before investing in the expensive HDTV equipment needed to produce the programs.[5] Although adoption of color television faced a similar chicken-egg dilemma, the demand for HDTVs might be slower because consumers on average may find that the additional service value from owning an HDTV set or VCR does not justify the cost of the new set.[6]

Another complicating factor to consider is whether households on average, regardless of programming, will want to pay for the extra computing power of an HDTV receiver. From a purely technical perspective, HDTVs and advanced computers could become substitutes. But the average TV viewer may not be willing to pay for a HDTV set that can double as a sophisticated advanced computer.

Standards. The timing and perhaps degree of consumer demand for HDTV will also depend on the selection of transmission standards. Before HDTV programming and sets can be widely introduced in the U.S. market, the FCC must set the standard that will define U.S. HDTV broadcasts. Over-the-air broadcasts are one of the primary methods to deliver the HDTV signal from the production studio to the home. The other three are cable, direct broadcast satellite (DBS), and fiber optics. Each involves a different approach to balancing the efficiency and quality of the HDTV signal. Selection of a transmission system requires weighing for each the technical merits, resolution improvements, spectrum requirements, impact on consumer cost, compatibility with the U.S. environment in the short run, and flexibility to address American video communication needs over the long term.

To date, HDTV in the United States has focused on terrestrial broadcasts, which account for 54–59 percent of television audiences.[7] Terrestrial broadcasters are also the only ones required to provide local services such as coverage of local news and elections.

Most of the transmission systems scheduled for FCC testing would require consumers to purchase new higher priced sets if they want to view the sharper HDTV images. Proposed intermediate steps would work with existing sets to enhance the conventional television signal. Alternative forms of HDTV signal delivery, such as fiber optics and the DBS approach, are also expected to play a role over the long term for the United States. Their role depends in part on finding a workable balance between the competing media such that competition and incentives to invest in networks such as the public fiber-optic highway are not discouraged.

For the potential signal deliveries, HDTV represents both the risk and promise of greater competition in the television industry. Satellite and cable firms are continuing to challenge the networks: HDTV represents an opportunity for them to further increase market share. Terrestrial broadcasters view HDTV as an opportunity to improve their ability to deliver better images and to limit competition with rival mediums. They must first demonstrate, however, that they are a viable medium for a full HDTV system. The competition for influence in the next generation of American television has already stimulated dramatic advances in compression techniques unimaginable even a year ago.[8] Another aspect of the competitive race involves the regulatory debate between cable television and local telephone companies. A central issue is whether restrictions should be lifted so that the telephone companies can carry television signals and perhaps facilitate the creation of a fiber-optic highway that may carry several HDTV channels to homes.

The technical, political, and economic challenges of HDTV have combined to slow the introduction and development of consumer HDTV in the United States—a slowdown that may prove advantageous (see the Appendix). Initially the Electronics Industry Association projected that "significant" HDTV sales would begin in 1993 if the FCC set a standard by 1991. Such a standard is not expected until mid-1993, with equipment sales lagging by one to two years. In the meantime production studios, receiver manufacturers, and consumers are waiting to see what the next generation of television will entail. On the downside a delay in standard selection may stimulate consumer confusion reminiscent of the VCR battle between VHS and Beta formats, encourage the emergence of expensive and complicated multiport receivers, increase upward pressure on the HDTV offer price, and further complicate the challenge of widespread household penetration. Yet the delay may prove worthwhile if it permits the development of a fully digital HDTV system that along with other high-definition systems is capable of ushering the country into the next information age.

The Critical Link in the Digital Revolution?

Much of the HDTV debate assumed that the industry's development and production would be the locomotive to drive U.S. firms in other electronics industries down the learning curve in the development of advanced digital technologies and systems and that government assistance was necessary. Contrary to the first assumption consumer HDTV is not expected to be the first major market or even the most important one for some of the technologies. More important, the need for the government to intervene to target and to elevate the consumer HDTV (household video entertainment) industry above other applications has not been demonstrated.

The linkages or interdependencies between consumer HDTV and other high-definition systems (present and emerging) are undeniable, but no consensus exists among technology experts on the central importance of consumer HDTVs for the information-processing and communications industries. Instead differences in regulatory structures and existing market conditions seem likely to limit the economic role of HDTV and the degree of commercial dependence of other industries on its domestic development.[9] In addition the competition between rivals in the television industry and between those in computers has generated vast improvements in some of the underlying technologies and components surprising even technology experts. These improvements throw into further doubt the need for government to give preferential treatment to domestic HDTV producers to ensure the development of high-definition technologies and systems. Clearly consumer HDTV is neither the only path forward for advancing sophisticated digital technologies and information systems nor the all-important one. Considerable investments and advances in the computer and communications industries are already playing a vital role that should not be discounted.

Perhaps the most telling judgment that HDTV is no more important than some other applications of the underlying technologies is the finding of the Office of Technology Assessment in a 1990 report on U.S. manufacturing and HDTV that "it is often impossible to be certain which application is ahead, or will remain ahead, as the driver of many of these important core technologies." OTA states further in the report that "the technological spillovers among different branches of the electronics industry cannot be pinned down or forecasted with precision."[10]

This central point means that support for "core technologies" cannot simply be equated with promotion of a single industry. Or in the case of HDTV, the development of the core information production, transmission, production, and display technologies cannot be reduced to the activist proposition that either the nation promotes consumer HDTV and captures the benefits of the technologies or it does not. If HDTV itself were the only application or the uncontested driver of the various technologies and components, then the distinction between the technologies and the application might be irrelevant. But as the next sections demonstrate, this is not the case. Rather HDTV will be central to some technologies, will benefit from some being developed first and reshaped in other applications, and will have a marginal impact on others. To accept the activist argument that consumer HDTV represents the link in the electronics food chain that would determine the commercial competitiveness of the other industries and their offspring would oversimplify both the nature of technological change and the digital revolution, which is drawing the various industries closer together.

Some industry analysts have rejected the claim of AEA and other activists that Japan's HDTV programs require the United States to jump in with an industrial policy to catch up with a program to revitalize the consumer electronics industry led by HDTV. George Gilder, with an alternative view of the digital revolution, agrees with the AEA that the line between television and computers is blurring and that future TV sets will be more like computers rather than simply better TV sets. He argues, however, that the analog-based HDTV system the Japanese have spent two decades developing is a real lemon that should be ignored rather than feared by the United States.

Gilder has emphasized instead the computer side of the electronics tree as the launching pad for developing television-computers and for leaping past the Japanese rather than simply catching up with them. The wave of the future, Gilder argues, is a telecomputer or a fully digital HDTV system. The advantages are more efficient use of broadcast spectrum from compression of video data into a series of ones and zeros and greater interoperability with computers. He finds, in contrast to the industrial policy advocates, that the United States is already moving toward telecomputers and the multimedia promises of digital information services based on U.S. strengths that have accumulated over time in computers, telecommunications, and some of the underlying components. Rather than industrial targeting, Gilder concludes that government's role should be limited to setting the parameters such as eliminating unnecessary regulations that discourage firms from tying together telecomputers with fiber optics or other advanced cables.[11]

Others have countered that Gilder is overly optimistic about the speed with which digital television–computers will be developed.[12] According to this argument digital television is a distant possibility, because of the need for dramatic advances in video compression and processing technologies. HDTV is necessary, the argument continues, as an intermediate step that policy needs to encourage, or else U.S.-based firms will fail to make the investments necessary to realize the telecomputer future depicted by Gilder.

Regardless of who proves to be correct, this debate among the technology experts amply illustrates that strategic industry promotion of HDTV is not a simple question of ratifying an industry consensus on which technological path the United States should follow. Rather, fundamental differences exist, and the government's decision would in effect select one version of the future, would discount the national opportunities offered by the other paths U.S.-based firms are currently pursuing, would shift resources more toward the HDTV path, and thereby would elevate the HDTV path above alternative paths. Furthermore, the decision is necessarily a speculative one because of the nature of spillover benefits that cannot be quantified and are spread over time.

Relative Size of the HDTV Market. Market volume and performance requirements are the two conditions most frequently cited for determining an industry's importance as a driver of the spillovers associated with the underlying technologies and components. Questions have been raised about the relative importance of consumer HDTV in both cases. HDTV's role as a technology driver is also unlikely to be independent of market size but seems instead to be integrally tied to its success as a high-volume, low-cost consumer product.[13] To examine the relative importance of HDTV, the Congressional Budget Office reviewed the three major forecasts of HDTV's market potential. Even after using optimistic penetration figures based on color TV and VCR adoption rates, the CBO found that the relative size and importance of HDTV as a consumer electronics product had been exaggerated. First, over the next twenty years HDTV receivers and VCRs are likely to remain a small portion of the electronics market. Although the AEA data suggest that in twenty years the global market for HDTV receivers and VCRs is expected to be $29 billion (in 1988 dollars), the $461 billion global electronics equipment market in 1988 already dwarfs this projection. The CBO also noted that annual growth in electronics is expected to be twice that of HDTV and the total consumer electronics sector share of the electronics sector is expected to remain roughly 20 percent.

Given the small scale of the consumer HDTV market relative to other applications, the CBO concluded "that it is counterintuitive to suggest that a small market that may exist in the future is more a driver of economies of scale, technology, and competitive success than is the growth in the present market."[14] This point is dramatically illustrated in figure 5–1, which contrasts the sales of HDTV with those of personal computers over the next twenty years, based on estimates from the American Electronics Association.

On the issue of performance requirements and HDTV as a driver of innovation, some questions have also been raised about the importance of additional HDTV research for other segments of the electronics industry. Engineers at Texas Instruments told the *Economist* that they are skeptical about the degree of commercial dependence between HDTV and chips since HDTV is not expected to be an independent determinant of success in chips. Instead future performance has been connected to the design skills and software required to make digital signal processing chips for a variety of final product markets, not just HDTV.[15] The next four sections discuss in more detail the questions that have been raised about the strategic industry role of consumer HDTV.

FIGURE 5–1

COMPARISON OF PROJECTED SALES FOR HIGH-DEFINITION TELEVISION
AND PERSONAL COMPUTERS, 1990–2010

Billions of 1988 dollars

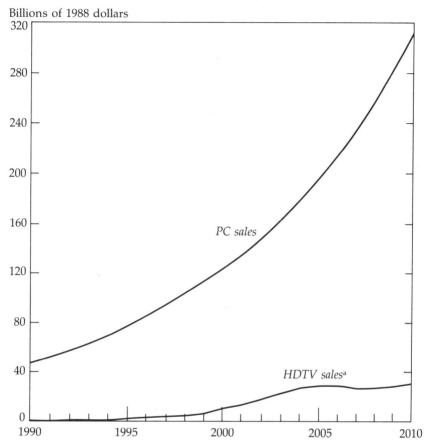

a. Includes both HDTV receivers and videocassette recorders.
SOURCE: American Electronics Association, "High Definition Television (HDTV):
Economic Analysis of Impact" (Report prepared by the Advanced Television
Task Force, Economic Impact Team of the American Electronics Association,
Santa Clara, Calif., November 1988), cited in U.S. Congressional Budget
Office, *Using R&D Consortia for Commercial Inovation: Sematech, X-ray Lithography, and High Resolution Systems* (Washington, D.C.: CBO, July 1990).

HDTV and Semiconductors. Frequently the case for HDTV is confused
with the strategic industry case for semiconductors.[16] If for the
moment the semiconductor argument is accepted, there are reasons
to resist the extension of this case to cover the consumer HDTV
industry and to question the nature of the link between the two

markets: is HDTV, for example, expected to be a primary source of demand for semiconductors?

After examining HDTV's role as a semiconductor demand driver, analysts have concluded that the large-scale market for semiconductor sales, which HDTV industry activists have listed as a cause for federal HDTV promotion, *already* exists in the growing computer industry.

Even under an optimistic scenario for HDTV sales (that consumers will be attracted to HDTV as rapidly as they were to color TVs) and a pessimistic outlook for semiconductor sales, HDTV would still account for only a small fraction of the aggregate semiconductor market in the foreseeable future.[17] Kenneth Flamm of the Brookings Institution has clearly demonstrated with industry data that HDTV's consumption of dynamic random access memory chips (DRAMs) is highly unlikely to surpass that of computers in the next twenty years. Indeed the trend in memory consumption appears to be away from consumer electronics products toward computers and office automation in Japan, one of the largest consumer electronics markets.[18]

HDTV and High-Resolution Displays. Some have also questioned whether the success of flat panels will hinge on the consumer HDTV industry. A common joke in the industry is that high-resolution flat panel displays have been just around the corner for at least twenty years. The HDTV industry seems an unlikely forum to speed their development for the foreseeable future. In contrast to home entertainment, the needs in business for high-resolution systems are immediate. The computer industry thus seems a more likely candidate to advance LCDs as computer manufacturers move to develop the high-resolution color displays to meet their customer demands rather than risk the wait for HDTV and its customers to emerge sometime over the next two decades.[19]

Given that household demand is expected to be sensitive to price, the cost of large displays must be reduced for HDTV to become a mass consumer product. As discussed in chapter 4, this reduction in turn requires breakthrough improvements in display and manufacturing technology. That will depend on precision manufacturing of a giant integrated circuit—a considerable technological challenge for engineers in the United States as well as those in Japan. The most common flat panels currently available are 10-inch monochrome displays. (Toshiba and IBM have hooked up to develop and perhaps to mass produce a 10-inch color active-matrix LCD by 1991.) The manufacturing puzzle of how to produce at economically efficient levels the large displays needed to make HDTV commercially viable is as a result not expected to be solved within the foreseeable future.[20]

In the meantime the different uses of displays outside the TV business create important incentives for a wide range of firms to develop improvements in non-CRT (cathode ray tubes) product and process technologies to satisfy demands of their customers. For computers the immediate challenge is delivering high-resolution digital video on medium-size screens. As a result, the use of displays is already growing in the industry, and computer firms are by some definitions already using higher-resolution displays than those proposed for HDTV.[21]

Because of poor timing, the potential importance of the consumer HDTV industry as display driver for the computer industry may be at least partially bypassed by developments in the 1990s. One issue is the speed with which display technology matures and the degree of learning that is necessary before the technological and manufacturing hurdles are overcome in the displays used in the computer industry. Dynamic learning effects in displays relevant to computers may not last long enough, or may be exploited by computer manufacturers before, HDTV becomes commercially viable. Some technology experts believe, for example, that the race in flat panels will be won by computers before television can even become a serious contender.[22]

In any event, if appropriation problems impede the development of display technologies common to a variety of industries, policies other than domestic promotion of HDTV production are needed. Rather than attempts to steer resources toward local HDTV production, research on display technology may in this context merit federal funding if some basic display principles still need to be explored or if the research is expected to have widespread applications and is not being funded at a desirable level.

HDTV and Advanced Computers. Although technical linkages will clearly exist between HDTV and computers, commercial success in computers has not been shown to depend on a domestic HDTV industry. The costs of HDTV to producers and consumers for conversion, the budding transmission technologies, and regulatory oversight are slowing consumer entertainment application of advanced imaging technologies. Computers offer a contrasting example: producers do not have to overcome the FCC regulatory hurdles, demand is not as sensitive to price, and market growth is rapid. The consumer HDTV industry, when it emerges, will likely help to lower the costs of large displays for business applications such as video-conferencing in addition to opening new markets, but there is no indication that the industry will determine the commercial competitiveness of computer manufacturers, let alone the economic health of the nation.

71

In terms of technologies the distinction between advanced televisions and advanced computers may seem irrelevant: TV sets like other consumer electronic products in the future, technically speaking, will be a computer on the inside. But whether the industries merge is a fundamentally different issue. The uses of TVs and computers in the home may continue to differ. Additionally, similarities in technology and economies of scope and scale are not sufficient for actors in one industry to become major players in the other. In computer and communication technologies, skeptics have pointed to the difficulty of large computer manufacturers such as Digital Equipment in entering telecommunications and of firms such as Northern Telecom in entering the computer field. Smaller computer network specialists were more successful in straddling the two markets. Big is thus not necessarily better: product specialization is as important as similarities in technologies for achieving success.[23]

Similarly HDTV activists largely ignore that when convergence in underlying technologies will translate into market convergence depends on an intricate mix of demand and supply factors that is largely unpredictable. James C. McKinney, chairman, U.S. Advanced Television Systems Committee, underscored this point when he warned a congressional subcommittee in 1989 that the enthusiasm about the prospects of new technologies and innovations such as HDTV needs to be tempered with a dose of pragmatism on the shape and pace at which the related final-product markets and services will develop. He noted that many scientists, impressed with the possibilities offered by technological advances, incorrectly predicted ten years ago that videotext would result in most Americans reading their newspapers on their television sets, while cable television promised interactive services for shopping, banking, and other high-tech functions.[24]

Improvements in technology clearly create new choices for consumers. But the economic question is the extent of market demand for the innovation: will TVs be used primarily for entertainment, and computers for design and desktop publishing? The use of separate rooms for entertainment and work may slow market convergence. The timing and degree of market convergence will also be influenced by factors such as consumer preference for a multipurpose set that has interactive capabilities, compatible video standards for television and computers, and the general regulatory framework—all of which are not yet known.

Workstations. Some have also criticized HDTV activists for exaggerating the strategic importance of HDTV for the workstation market.

The high-paced workstation market is expected to grow 46 percent in 1990 and by some estimates is at the forefront in computing and display technologies. The global market is dominated by four American firms—Sun Microsystems, Apollo, Digital Equipment, and Hewlett-Packard—that together accounted for 75 percent of worldwide shipments in 1987.[25]

The engineering community has been the primary user of workstations with an emphasis on software development and computer-aided design as in integrated circuit design. The system components include high-resolution color displays, fast processors, massive memory, special application programs, and local area networks that link users. In recent years strong competition between the four top workstation firms and between four major PC manufacturers—IBM, Apple, Compaq, and Data General—has stimulated dramatic improvements in equipment performance, accelerated the pace of new product introductions, and put downward pressure on price. Accordingly, analysts expect growth in workstations to remain strong and coverage to expand to include commercial markets in areas such as electronic publishing, office automation, and image processing.[26]

The development of a domestic HDTV industry has not been considered a vital ingredient for this growth for several reasons. Application-specific problems and the needs of workstation users are expected for the foreseeable future to differ from those of prospective household HDTV consumers. At a public hearing on HDTV, the Electronic Industry Association pointed out that workstation displays do not need to be as bright and large as those required to make HDTV sets attractive to household consumers since engineers tend to huddle over their displays while TV viewers are typically seated several feet from the receiver. Workstations are also expected to surpass by a significant margin the processing requirements necessary for a consumer electronics product.[27]

Moreover the introduction and development of high-definition technologies in the workstation environment may be helped by a largely untapped market. The International Trade Administration has estimated that only 3 percent of the potential workstation global market has been penetrated, while demand is mushrooming.[28] In contrast, consumer HDTV faces a saturated conventional television marketplace where 98 percent of U.S. households already own a TV set. If HDTV fails to alter consumer buying habits dramatically and if they continue on average to hold onto their sets for seven years, then the HDTV rate of household penetration will depend on this replacement rate, which will limit the speed at which the consumer HDTV market evolves.

Trying to speed the convergence of television and computer markets may furthermore yield disappointing commercial results. The hallmarks of the consumer electronics industry have been video and sound delivered at the lowest possible cost. In particular the key to success in the television business has been not technological sophistication but the decision of a relatively large number of consumers to buy the product at a profitable price for producers. With this history relevant questions have been raised about the wisdom of building an HDTV set that tries to do too much and includes too many technological extras. Technological virtuosity may not be synonymous with commercial success.[29]

Multimedia applications. If high-resolution systems and advanced imaging technologies are expected to be the key to the next information age, what is the private sector doing about it? Promotion of consumer HDTV as a strategic industry is not simply an issue of identifying the technical possibilities and then starting anew with an industry toward this future. Rather, this policy approach requires a choice between this and the other paths already being pursued by the private sector.

Before a reasonable decision can be made about the need to redirect the evolving digital electronics marketplace toward HDTV and consumer electronics, several basic decisions are necessary. These include whether the research and production plans of computer manufacturers are shortsighted with respect to national technological opportunities for growth, what the dynamic implications of their actions are for the U.S. national economy, and most important whether government intervention to favor HDTV production—which may raise the resource costs of computer manufacturers today—will promote long-term growth opportunities for the national economy. Despite their central importance, these issues were given scant consideration in congressional hearings on HDTV industry promotion plans.

It helps to start with the basics. The market players in the $300 billion global computer market interested in the creation of an interactive computer or Guilder's telecomputer are not waiting for policy makers to target the HDTV industry with pledges of taxpayer support.[30] They have been unwilling to wait for a $10 to 20 billion American HDTV market to develop over the next twenty years.[31] In an industry where complacency can be fatal and success may be cumulative, they are already investing in the digital technologies and applications that they expect to shape the global computer market well into the twenty-first century.

The positioning of world computer and consumer electronics companies has spawned an array of corporate alliances to develop multimedia products. These multimedia products involve the combination of video, graphics, audio, text, and numbers on a computer. Efforts by the computer industry to develop these products appear poised to contribute fundamentally to the convergence of computer and television technologies. The multimedia system uses a high-definition color computer monitor and special chips to mix video images, taken either from a laser disk or off the air, with information stored in computer memory. Although this is not a novel concept, advances in chip technology have improved image quality and the speed of integration and focused attention on multimedia as a central force of change in the computing environment for at least the next decade.

The video component has been the main technical stumbling block in the multimedia path because of the vast volume of data that need to be translated into a digital code accessible to computer manipulation. Research and development efforts have revolved around compression techniques that minimize the space needed to transmit and store video data without significantly compromising image quality. Recent developments in this area have been characterized as a bridge between sophisticated TV images and the crude graphics of video images typical of earlier interactive video systems.[32] Based on these developments computers can treat moving video images like a spreadsheet and for the first time obey the grammar of television, which permits the blending and fading of images. One application may be a student interacting with a National Geographic film on monkeys via an electronic pointer to touch images and access information stored in a computer data bank. Another is a research project that allows the student to do video research and create a presentation that shows, as well as tells about, the historic moments in the life of figures such as Martin Luther King. For business the multimedia age involves the transformation of personal computers into multipurpose interactive workstations.

The Information Workstation Group of Alexandria, Virginia, projects that the worldwide multimedia market will grow from the current $6.4 billion to $24.1 billion by 1994. Lured by this prospect, IBM and Intel have joined forces on one multimedia path to develop video- and sound-processing technology to be used in personal computers. The centerpiece of Intel's approach is digital video interactive (DVI) technology, developed at the David Sarnoff Research Center; this is a set of chips designed to convert PCs into multimedia workstations. Other components of Intel's approach include an alliance with Olivetti to adapt DVI for Europe and with Thorn EMI to market DVI in the United Kingdom. Apple has also decided to target the business market and is investing in the development of a multimedia computer based on the Macintosh PC that is expected to be introduced within the next few years. Sun Microsystems has recently joined forces with a semiconductor manufacturer, Texas Instruments, and a leading consumer electronics research facility, the David Sarnoff Research Center, to develop a multimedia workstation, which may lead to innovation in computer architecture.[33] Philips has emphasized the consumer end of the market with a strategy based on compact disc interactive (CD-I) technology and the support of two consumer electronics giants, Sony and Matsushita, and Intel's arch rival, Motorola. (For a roadmap of the race see figure 5–2.)

The variety of investment strategies indicates that the multimedia and high-definition system race has already started. Consumer HDTV is only one of the many systems involved. The U.S. market is an integral part of the race, and U.S. firms are positioning themselves to take advantage of potential markets in the home, business, as well as education. The range of investment plans reflects the range of opinions on the best strategy for the race. In this context, it is impossible to conclude that those who have chosen to focus on consumer HDTV should be favored.

Moreover the multimedia race illustrates the complexity and inherent difficulty of an industry-specific approach to supporting the commercial development of advanced technologies: often multiple investment strategies and paths exist, the competition between participants is fierce, the pace of change rapid, and the nature of change difficult to anticipate. Under these conditions even well-informed bureaucrats would have difficulty determining the extent to which market participants are underinvesting in the development of a particular industry and thus the amount of assistance required to improve the allocation of the nation's resources and the nation's technological capabilities. They would have to answer questions such as why U.S. production of advanced televisions is more important than workstations, whether domestic production should be supported with direct production subsidies and if so, how much, and whether import competition should be restricted and if so, how and for how long.

Further, in many respects the industrial policy debate is a costly diversion of national attention from pressing matters related to commercialization of advanced technologies relevant to, but not unique to, a particular industry. The HDTV policy debate has eclipsed the troubling issue of compatibility between the various information systems now emerging. If compatible standards develop in the various high-definition systems that are likely to become available to house-

FIGURE 5–2

COMPUTER MANUFACTURERS AND
THE MULTIMEDIA RACE

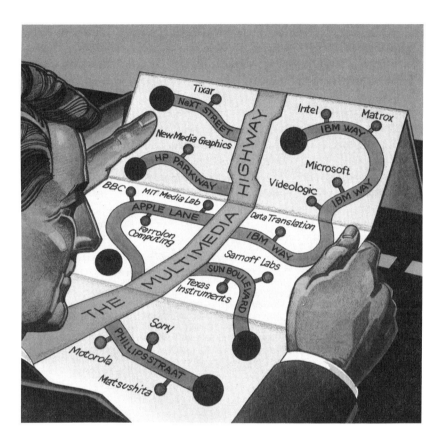

Reprinted with permission from *Electronics* Magazine, copyright © February 1990, Penton Publishing, Inc.

holds and businesses, economic opportunities can increase. Future homes are likely to have television, computers, and video phones as the major platforms for information services. If the three can exchange information in a cost-effective manner, consumers may find additional service value in the products.

Thus far, the standards process has largely been dominated by broadcasters and consumer electronics manufacturers with little involvement of computer manufacturers. At the center of attention are the technical parameters (resolution or pixels per line, square or oblong pixels, and frame rate) that will determine the extent and ease with which television and computers will be able to exchange information. Recent moves by Zenith to alter its proposed production standard and those by other HDTV participants to develop fully digital HDTV transmission systems have helped to open the door to compatible standards.[34] Ensuring that similar moves are not discouraged is a more concrete way to support U.S. interests in exploiting the commercial potential of cross-cutting technologies than speculating about which current or emerging application merits federal favor. (See the Appendix.)

HDTV and Telecommunications. Equally important are the barriers that may curtail the infrastructure investment needed to link public telephone, communications, computer, and television technologies and applications. The path toward this objective is at present unclear.

Fiber-optic wire has been tapped by a number of analysts as the future route for widespread delivery of information services. Like an interstate highway system that permits the physical delivery of goods and services, a fiber-optic system would move electronic information. Such a system would increase reliability, speed, and capacity of the network to carry information relative to current phone lines and cables. Much of the uncertainty over the creation of this highway has revolved around questions of timing: do applications need to come first to stimulate the necessary infrastructure investment? Are the advanced technology products emerging from the computer and telecommunications industries insufficient to encourage this investment? What regulatory, economic, and political barriers are inhibiting investment?

Regulatory concerns have focused on the extent to which the investment incentives of regional Bell operating companies may have unnecessarily been curtailed. The Bell companies have argued that they need to be able to provide information services including television to justify their investment in a fiber-optic network. Some advocate removing any constraints on competition between these companies and cable operators to encourage the competitive race between these two groups to connect homes to fiber networks.[35]

Other concerns stem from the different agendas of industries with a stake in the delivery of future information, communication, and entertainment services. Without an accepted common objective the rate at which an efficient and effective public data highway is built may be slower than socially desired. This raises the prospect of a network externality problem and underinvestment in infrastructure:

compatibility or interoperability in information and communications systems may prove to be elusive if left to market forces and may therefore merit government attention. Attention to these compatibility and network issues would serve national economic interests in developing the underlying set of technologies better than an endless debate over the philosophical merits of industrial policy.[36]

What HDTV Threat from Abroad?

In addition to exaggerating the importance of consumer HDTV, the threat to national competitiveness from foreign HDTV research programs has been overstated. It is true that the Europeans and Japanese are ahead in introducing an HDTV system and expensive sets (single sets cost between $18,000 and $34,000 in Japan) to the market.[37] But they are also in the process of backtracking and second-guessing their HDTV plans.

The HDTV system that the Japanese have spent two decades developing is based on conventional TV (analog), which engineers in U.S. corporate labs have already bypassed in the competition to develop a digital HDTV system. By targeting and investing in a hybrid analog-digital system, Japan gambled that the experts would be right and that a fully digital HDTV transmission system would not even be a possibility until some time in the next century. But recent advances in compression and digital signal-processing technologies suggest full digital transmissions are possible, placing the Japanese in the unexpected position of trying to catch up to the United States, a latecomer to the HDTV race, in digital video transmission.

The Europeans face a more difficult problem. Heavy research subsidies to develop their analog-based HDTV system HD-MAC over the past six years has succeeded in producing a system that few (Europeans included) are willing to adopt. The unforeseen U.S. advances in digital compression technologies are further forcing European governments to confront the possibility that their HDTV plans (HD-MAC) missed the mark and may already be outdated.[38]

Ironically, although the United States chose not to heed the advice of activists to map out a national HDTV plan and to underwrite it with a major investment program comparable to European and Japanese programs, the United States may yet emerge the winner, with a superior HDTV system more adept at communicating with other digital electronic systems in twenty-first century homes and businesses. Moreover the recent technological leaps made by Americans underscore the inherent difficulty of forecasting the future especially in dynamic, high-tech industries.

79

A Global Perspective

Considering how HDTV fits into a rapidly changing global economy helps put the strategic industry issue into perspective. The framework consists of an expanding web of research and production networks that cut across national borders and thereby erode whatever edge a national industrial policy may have had. As the importance of multinational corporations and international interfirm alliances increases, the ability of governments to target national technological capabilities through industrial subsidies declines: government research support for domestic firms may, for example, have little affect on the location of HDTV production.

Conversely, as discussed in chapter 2, the growth of international alliances and advances in communications technologies have increased the likelihood that HDTV-generated technological spillovers will be international in scope. The United States may as a result appropriate some of the benefits generated by targeting practices abroad, undercutting the case for a defensive industrial policy to counter foreign HDTV research programs. Put simply, there is no indication that strategic industrial targeting can significantly alter a nation's ability to appropriate technological capability in a world marked by multinationals, a growing web of cross-border partnerships, and geographically dispersed industries.

International Technology Networks. Cross-national partnerships have become a key feature in the latest stage of global integration. Rather than the capital flows characteristic of previous stages, the exchange of technological information and design capabilities and managerial expertise between multinational corporations are central features of the current stage.[39] Advances in transportation and communications have increased the speed of technology diffusion and reduced the cost of such transnational communication. The transaction costs of transferring data to partners in a joint venture or working with subsidiaries abroad have fallen, for example, because of enhanced digital communications networks. With the increase in interfirm cooperation, firms are also no longer limited to bargaining with each other through price or traditional market relationships.

During the 1980s transnational corporate alliances proliferated, expanding the range of technology development options and quickening the pace of international technology transfers. Mature, middle-aged, and emerging industries are all on the list of activities where joint ventures became important. Steel, automobiles, aircraft, semiconductors, and increasingly consumer electronics are particularly

TABLE 5–1

THE U.S.-JAPANESE CHIP CONNECTION SINCE 1986

U.S. Firm	Japanese Firm	Year
Equipment Manufacturing Ties		
Intel	Matsushita	1988
	Mitsubishi	1987
Motorola	Toshiba	1987
National Semiconductor	Mitsubishi	1987
Powerex	Mitsubishi	1988
Signetics	Matsushita	1988
Texas Instruments	Mitsubishi	1988
Weitech	Matsushita	1988
Technical Exchanges		
Intel	Fujitsu	1986
LSI Logic	Sharp	1987
MIPS	NEC	1989
AMD	Sony	1987
Motorola	Toshiba	1987
Texas Instruments	Hitachi	1989
	Sony	1988
VLSI Technology	Sanyo	1988
VTI	Hitachi	1988
Joint Venture		
Motorola	Toshiba	1988
Product and Production Sharing		
Intel	Mitsubishi	1987
National Semiconductor	Mitsubishi	1988
Motorola	Fujitsu	1987
	Toshiba	1987
	Mitsubishi	1987
Texas Instruments	Fujitsu	1987
	NEC	1988
	Mitsubishi	1986
Other	Fujitsu	1988

SOURCE: In-Stat Inc., in Jonathan Weber, "U.S., Japanese Firms Find a Common Cause," *Los Angeles Times*, July 2, 1989, p. 1.

prominent examples. In the semiconductor industry alone 121 international technical agreements were signed by forty-one firms in the United States, Europe, Japan, and Korea between 1978 and 1984 (for example, see table 5–1).

The international alliances and licensing agreements that have already been formed in HDTV and the linked semiconductor and display areas underscore the trend toward the international organization of research and production (table 5–2). Two of the major actors in European-sponsored HDTV research have joined forces with the American consumer electronics research group, the David Sarnoff Research Center, to develop a simulcast digital television system in the United States. IBM and Texas Instruments are diligently tapping Japanese experience to carve out their market shares in HDTV and other high-definition systems. Equally important, the interindustry alliances (table 5–2) also illustrate the multiple interests and strategies private firms are pursuing to internalize the information flows they consider necessary to develop high-definition systems and promote their competitive interests. The AT&T alliance with Zenith reflects, for example, their interest in a high-definition system tied to the telecommunications network.

Semiconductors, displays, and HDTV are pieces of the much larger trend toward global markets that are reshaping the structure and strategies of both firms and nations. Although opinions differ over the breadth and pace of current moves toward transnational actors and global industries, the direction is unmistakable and largely irreversible.[40] The need to share the rising costs of advanced R&D and to tap growing markets abroad has pressured firms to develop global strategies and experiment with cooperative alliances. According to ITC Chairman Anne Brunsdale in 1989, "globalized planning is now itself a factor of production, and sales and profits will go to the firm, regardless of its nationality, that has comparative advantage in its use of the world factory."[41]

As firms continue to compete and to cooperate in the race to recoup their investments and to learn from each other, nations will increasingly be forced to compete for the domestic location of investment. The tools of industrial policy are poorly designed for this competition for several reasons. They may subsidize the creation of proprietary knowledge or difficult-to-copy know-how embodied in the organization of a firm, but these firm-specific assets seem less constrained to the domestic environment than in the past. Spillovers sponsored with public funds may stay with firms but leave the country as firms trade industrial assets and organize globally to exploit opportunities for market share and technological advance. The United States may thus gain only marginal benefits in the long run if the firms singled out for HDTV research subsidies decide to make televisions offshore in lower-wage countries or to use foreign inputs. This possibility raises a fundamental question about a basic industrial

TABLE 5–2

SELECTED ALLIANCES IN HIGH-DEFINITION SYSTEMS AND COMPONENTS

Firms	Objectives
Texas Instruments, NHK [a]	Technology transfer, license for the design of semiconductors used in decoding for HDTV sets from NHK to TI
Texas Instruments, Sony	Use by Sony of TI's digital signal processors in new CD players
IBM, Toshiba	Joint venture to produce LCDs for next generation workstations and computers in Japan; also IBM interest in manufacturing technology from Toshiba
Xerox, Standish	Joint development agreement to explore possible manufacturing and marketing of high-resolution active-matrix displays
NBC, Philips, Thomson, David Sarnoff Research Center (DSRC)	Introduction of enhanced-resolution TV and joint development of a digital HDTV simulcast system for the United States
Zenith, AT&T	Joint development of fully digital HDTV transmission system for the United States (including HDTV receiver)
General Instrument, MIT	Joint development of fully digital HDTV transmission system
Thomson, Philips	Estimated $3.6 billion joint development of HDTV sets by 1995, including cooperation in flat screens, integrated circuits, studio equipment, and final sets
Eureka (EU95)	Creation and demonstration of European HDTV based on MAC family of standards, with participating firms from the Netherlands, France, Germany, Italy, United Kingdom, Belgium, Sweden, Finland, and Spain
Sun, DSRC, Texas Instruments	Joint development of high-definition video workstation (some funding supplied by DARPA as part of high-definition display project)

a. Japanese broadcast company.

policy assumption that promotion of potential HDTV producers would generate major technological benefits for the national economy.

International Trade in Components. Consider again the example of the Swiss watch industry, in which the economies of scale implicit in watch production depended on the magnitude of total watch demand worldwide rather than the scale of national output. A similar scenario seems likely for HDTV equipment. To the extent that the advanced components are traded, potential cost savings from expanded production of these products will depend on the magnitude of worldwide demand to produce final products such as HDTVs or advanced computers.[42] In this context the United States is unlikely to miss out on HDTV because of foreign HDTV research or production.

If HDTV is viewed as one element in the set of high-definition systems, then the status of U.S. participation looks less dire than the one painted by HDTV champions. The various strengths of American and foreign manufacturers suggest that specialization will continue to play an important role in the development of high-definition systems. Memory chip and display production are areas of Japanese strength, and areas of U.S. strength include digital video compression, software, system integration, microprocessors and their design,[43] logic chips, a competitive broadcasting industry, and the general advantages associated with the world's largest integrated electronics market.[44]

Discussions in progress between members of the U.S. Semiconductor Trade Association and the Electronics Industry Association of Japan, which aim to create an interface between U.S. chip suppliers and Japanese consumer electronics firms, also suggest cause for some optimism. U.S. semiconductor producers are expected to play a role in the development of consumer HDTV in Japan, particularly in the area of logic circuits.[45] Wilfred Corrigan, SIA chairman, also told *Electronic News* that "U.S. manufacturers are now starting to get a larger share of (product) design-ins in the Japanese market. This should lead to much larger production orders" that will become evident in the 1992–1993 period.[46]

The implications of this background for HDTV development and American technological capabilities are twofold: first, foreign HDTV programs do not necessarily hurt the United States and may instead increase demand for U.S. products, and, second, U.S. domestic HDTV promotion plans may not help much. Indeed the growth of overseas production and joint venturing is blurring the meaning of national technological capability.

Two cases in particular illustrate this point. One involves the proposed takeover by Fujitsu of ICL, the largest British computer manufacturer and an active member in the Jessi, Esprit, and Eureka programs. The second involves the proposed purchase of the American semiconductor equipment manufacturer Semi-Gas Systems, a supplier to the American private-public Sematech consortium, by Nippon Sanso KK of Japan.[47] Both define the electronics industry in terms of nationally owned companies that the projects try to bolster in an effort to improve national economic performance. Both reveal, however, the flaws of inward-looking policies and the illusory nature of indigenous technologies when capital is internationally mobile. Jessi, the $5 billion EC-funded microchip research program, was not intended to support Japanese firms, but a frustrated program coordinator and other members are now faced with exactly this situation. They must decide whether to expel ICL or adjust their original objective to reflect better an industry that is internationalizing despite their best efforts. Similar problems are also plaguing Sematech managers and member firms.

In addition, neither the innovation of HDTV nor foreign government HDTV research programs are likely to alter fundamentally the economics of location in the television set industry. Contrary to popular belief, as discussed in chapter 3, most of the sets sold in the United States are made here, and American workers add more than 70 percent to the value of the set. This may change, but the costs of transporting the bulky and fragile cathode ray tubes and advances in display technology are likely to play the key role rather than the nationality of who owns the HDTV technology or the plant: part of the display production that now occurs in the United States because of transportation costs may be shifted to less-expensive production environments either by Americans or foreign-owned firms.

Furthermore, rather than a positive effect, the result from U.S. HDTV industry promotion could be negative—for the industry as well as the country. Some manufacturers in the television businesses worry that this form of intervention could take a healthy outlook for HDTV and make it worse. In a congressional hearing on the AEA's industry-led plan for an ATV corporation, Philips expressed concern that such a plan would "delay rather than expedite U.S. progress on HDTV" and argued that the ATV corporation was unnecessary.[48]

Pushing a domestic HDTV industry in an international marketplace may also hurt rather than help the linked industries. The export success of the U.S. TV equipment business, which is dominated by the American firm Ampex, could be undermined if it lost valuable labor resources to HDTV or if the United States imposed import restrictions on HDTV inputs to encourage U.S. suppliers and then other countries retaliated. With import restrictions, other emerging downstream industries such as interactive PCs would be forced to pay higher prices for a critical input and international cost-competitiveness could be undercut. If trade in intermediate goods continues to expand and interdependence continues to increase, it will be even more difficult to protect and promote a domestic industry without hurting a linked industry. The next four sections highlight some of the practical problems illustrated by HDTV of the strategic industry promotion approach.

Would Domestic HDTV Producers Buy Domestic? Consider again a critical premise behind the push for HDTV promotion that large-scale domestic HDTV production would stimulate significant new sales for U.S. semiconductor producers. But would domestic producers buy domestic?

Although the stress of industrial policy advocates has clearly been ownership and location, neither saved U.S. component suppliers in the past. Even when American-owned firms dominated the U.S. consumer electronics market, components were imported.[49] Given that competition is expected to remain fierce in consumer electronics, domestic production of HDTV seems no more likely to ensure a safe market for U.S. component suppliers than in the past.[50]

Who Is Domestic? The expanding web of cross-border alliances and production ventures has sufficiently blurred national corporate identities such that industrial policy efforts at separating "us" from "them" are often reduced to exercises in futility and frustration. Who are "we" in the television industry? Ownership is no longer a reliable guide since the domestic industry is dominated by Philips and Thomson. As discussed in chapter 3, they have demonstrated a clear commitment in terms of research and production to the U.S. environment, are currently linked to HDTV efforts both abroad and in the United States, and are expected to have the edge in a number of the interconnected HDTV technologies. As a result, they are at least as American as Zenith. In a report to the National Center for Manufacturing Sciences, Jeffrey Hart noted that Zenith does not source from U.S. firms but rather primarily from ITT (Europe) and Matsushita. Thomson-USA sources from Motorola, National Semiconductor, and Harris (which purchased the semiconductor division of RCA) in addition to ITT and OKI, which also service Thomson-Europe.[51]

Who are we in television tubes? The challenge of identifying "who we are" is illustrated further by the previously discussed television tube case before the Commerce Department. Five Japanese firms and two Korean firms stand accused of circumventing a government order intended to ensure fair trade. The charge is dumping Asian tubes in Mexico, which are then used in sets sold in the United States market, allegedly causing material injury to the U.S. industry. A request has been made to apply dumping margins to the tube value of imported sets.

But who represents the industry, and who is injuring whom? Zenith uses Asian tubes in the sets it imports. Both Philips and Matsushita assemble sets in the United States with U.S. tubes and import smaller sets from Mexico. Yet Matsushita is accused of unfair trade in television sets. An additional twist is that economies of scale in production of tubes limits firms from sourcing components in every market they serve. The potential thus arises for the misuse of U.S. trade laws by one set of companies to attack their rivals because they organize their global operations differently.[52] Policy makers are left with the vexing task of deciphering who may be injured at the domestic level and when dumping margins can be applied to different products, while pursuing at the international level guidelines for managing circumvention problems in an era of flexible sourcing arrangements.

The case also signifies the advent of a new era in antidumping cases. More cases are likely in which foreign-owned companies with factories in the United States seek to use U.S. laws to their advantage by seeking protection from imports of other foreign-owned companies as well as from U.S. owned companies.

Who are we in displays? In addition to the considerable difficulties of forecasting technological relationships, internationalizing industries present new and perplexing questions on the relationship between downstream and upstream industries and the national interest. This is amply illustrated by an antidumping case filed by seven U.S. flat panel display producers against twelve Japanese manufacturers of the high-information displays. American computer manufacturers that use the displays have opposed the petition and have argued that there were few alternatives given the scarcity of viable domestic substitutes particularly in passive matrix LCDs, which are not manufactured domestically and compose the largest share of imported displays.

Among other issues the ITC and the Commerce Department in this case must balance the interests of U.S.-owned producers and employees against those of downstream users.[53] There is also no guarantee that antidumping duties would improve the outlook of American upstream display firms if American computer manufacturers decided to shift production of products such as lap tops offshore in pursuit of lower costs.

Can Government Do the Right Thing?

Finally, and most important, the HDTV debate has amply illustrated the hazards of high tech politics and industry-led targeting. Rather than one industry voice to guide policy makers, there are many. Just as multiple technologies are involved with HDTV and the related final-product markets, there are multiple interest groups—each with its own policy agenda and report documenting their critical importance. The range of interested parties includes TV programmers, TV equipment suppliers, cable companies, satellite companies, over-the-air broadcasters, domestic and foreign TV manufacturers, U.S. semiconductor manufacturers, telecommunications companies, and computer manufacturers—the list seems endless.[54] Which industry voice should be followed? Whose agenda for the development of high-resolution systems, standards, and components should be chosen? Who should decide?

Even the most well-intentioned in this environment would have difficulty in selecting winners and determining an economically efficient allocation of private and public resources. U.S. politicians will also find it difficult to treat industries solely on the basis of technical criteria and to avoid the political pressures generated by the "me too" syndrome. Jagdish Bhagwati provides an apt summary of the policy process with his conclusion that only in one's imagination is the U.S. government a puppet of economists (or, in the HDTV case, engineers), speaking with their voice and acting at their command to manage sophisticated intervention in the economy.[55] Limiting the scope of the "me too" effect seems particularly problematic when, as with HDTV, high-profile industrial policy tools such as grants or subsidies, import protection, preferential financing, and selective procurement have been suggested.

The Risk of HDTV Industry Promotion. The cost imposed on other activities in the U.S. economy by promotion of consumer HDTV and the trading system in general is a final, often overlooked set of factors. Whatever benefits the HDTV industry produces will not be free. Not only are there the direct taxpayer costs of research and production subsidies, but favored industries also inevitably draw resources away from other activities in the economy.

It is incumbent upon those who would change the current allocation in the name of improving national welfare to analyze the indirect costs imposed on other industries. Proposals such as the Consumer Electronics Capital Corporation (CECC) described in chapter 4, which are designed to provide consumer electronics producers with preferential access to capital, should be questioned about the effect on the short- and long-term cost of capital for other U.S. manufacturers. Government pledges of support and low-cost loans are intended to shift capital toward consumer electronics and to lower the cost of capital for this industry, but the cost of capital for other market participants may rise unless the pool of capital is somehow expanded by the program.

The costs from diverting scarce technical talent should also not be overlooked and should be weighed carefully against whatever static or dynamic benefits the HDTV promotion program may generate. Some have argued that a strategic industry plan for HDTV would secure high value-added jobs for U.S. workers. To validate this point, however, they must show that jobs would be created rather than shifted from other industries. Policy makers must ask where the workers and engineers will come from and at what cost to other segments of the electronics industry.

The claims that a strategic industry plan for HDTV would create jobs or secure domestic production needs to be evaluated not only in terms of costs to other industries but also in terms of alternative and more general job creation and infrastructure proposals. In particular the uncertain national benefits of spillover from favoring HDTV need to be weighed against the national returns from investment in transportation and communications networks and human capital, which have established records of payoff for the nation without the diversion costs and risks associated with industry-specific assistance.

Additionally the nation's commitment to the industry's development over time may entail other costs: consumer welfare losses from higher prices if trade is restricted, as some have proposed,[56] or efficiency loses from higher production costs, at least in the short run, from the higher cost of domestic production that might otherwise have occurred overseas. If the industry does not become competitive internationally, then the costs would further require adjustment to take into account a longer-term promotion and protection plan.

Finally, if everyone targets HDTV, who wins and at what cost to the international trading system? If the United States and its trading partners simultaneously target HDTV and other industries to protect themselves against missing out on what another country has labeled as strategic, then the loser may ultimately be the rules-based trading system as it struggles to cope with policy induced trade tensions and problems of global excess capacity. This happened with steel and with semiconductors. HDTV may be next.

In summary, the HDTV debate demonstrated the inherent complexity, limitations, and hazards of the strategic industrial policy approach to promoting national technological competitiveness. In the high-tech sector in particular, the debate confirms that the ultimate (product) shape of new technologies will depend on a complex mixture of demand and supply factors, some of which are unknown or also depend on technological advances. As a result it is impossible to conclude that not enough of the nation's resources are being invested in the emerging consumer HDTV industry. Will HDTV or telecomputers come to homes first and which will be most important for the nation in the future are questions that can not be answered today. Nor is there any need to answer them: in addition to the need to focus on the general factors that define an attractive U.S. production environment, the government has a distinctive responsibility to fix only the specific things that are known to be broken—and the emerging HDTV industry is not. Furthermore there is no indication that the United States would miss out on HDTV because of targeted industrial research programs abroad. It is therefore impossible to conclude that reallocating private and public resources to favor the creation of a domestic HDTV industry would be welfare-enhancing.

Consumer HDTV may become an important industry. But that is not the industrial policy issue. The issues are whether a domestic base in consumer HDTV production is essential for future national income growth, whether and how much firms are underinvesting in the technological opportunities of the development of the underlying technologies, and what the government can do to improve the market results. The debate over HDTV has made clear that industrial policy initiatives are unwarranted and a different approach is needed to address legitimate national concerns in areas such as the expansion of markets worldwide and the commercial development of advanced technologies that cut across multiple applications. These concerns are not unique to the consumer HDTV industry or any one industry (existing or emerging) and industrial policy is therefore not a solution. But it is part of the problem.

6
Policy Implications

Over the past two years widely divergent definitions, agendas, and national economic concerns have become entangled in the HDTV debate. In many respects HDTV has become a test case for the larger debate over whether the United States should in the 1990s tilt toward a strategic industrial policy approach to support the development of advanced technologies.

To explore the particular challenge of HDTV activists, the beginning of this study asked two questions: is HDTV a strategic industry in economic terms, and is strategic industry promotion necessary? The first issue remains at best an open-ended question among technology experts and even if there were a consensus on the strategic importance of HDTV there is little evidence to suggest that strategic industrial targeting would be a cost-effective approach for the nation. The HDTV debate has furthermore demonstrated that the mere possibility of strategic industry promotion distorts incentives by encouraging firms and trade associations to lobby for special treatment on the ground that they, like previously favored groups such as semiconductors, are critical for the nation. Although consumer HDTV will generate spillovers for the linked industries, its importance as a strategic industry has been overstated and the threat of foreign targeting exaggerated.

The government has a distinctive role to play in establishing a general investment environment that will encourage the development and diffusion of advanced digital technologies while firms determine how to shape the technologies into advanced products such as HDTV, telecomputers, and other types of high-definition systems. Policy makers do not have to accept the conclusions of industrial policy advocates to recognize that some of their key concerns merit attention to issues such as standards, research funding for advanced technologies, and differences in industrial structures and practices abroad that may impede market access and distort trade.

91

Standards

The selection of an HDTV transmission standard should be timely—neither too slow nor too fast. This will be difficult because of the thin line between the inevitable positive and negative implications of the standards process. On the positive side a standard adopted early will reduce consumer and producer uncertainty, will lower transaction costs, and may permit the market to mature more rapidly. On the down side a standard adopted prematurely can freeze technology at a suboptimal level, can limit exchanges with other applications, can reduce the room for innovation, and can curtail market potential.

The Federal Communications Commission is charged with the technically and politically complicated task of spectrum allocation and the selection of television transmission standards in the United States. To date the FCC has made two major decisions. In 1988 the commission required that HDTV broadcast programs must be compatible with the existing stock of conventional NTSC sets. Household consumers will, as a result, be able to keep their conventional sets once HDTV broadcasts start, but they will not get the sharper HDTV picture unless they purchase a more expensive HDTV set. A timetable for the selection process was announced later (in March 1990) with the FCC's intention to adopt a digital standard, while not precluding intermediate options that the market and the current state of know-how on digital transmission may require.

As many of the high-definition technologies and systems are still evolving, a flexible approach toward standards should be encouraged across the board. This should explicitly recognize that high-definition television is not the only, or even the most important, emerging high-definition system. Converging computer and television video technologies open the door for new services and products in the home, business, production, and research environments. The government should be careful not to close this door. In particular, U.S. standards should recognize that compatible standards for video in television and computers can release equipment from a single application and encourage widespread diffusion of new technologies and services. Policy should also address unnecessary regulatory barriers that impede healthy competition and that may limit the market's ability to promote compatibility between emerging high-definition applications in television, computers, and communications.

Inattention to some of these issues in Europe and Japan ironically may make the United States as the latecomer the real winner in developing quality HDTV and high-definition systems for the United States. In particular, analog-based HDTV systems introduced in Japan and Europe do not resolve the video compression problem (that is the problem of "fitting an elephant into a bath tub," discussed in Chapter 4) and instead consume large amounts of limited spectrum and bandwidth. Analog will also make it more difficult to interface computers and other advanced digital applications that may develop.[1] (See Appendix for more details.)

Technology Research and Development

In a world where technological spillovers increasingly refuse to honor national borders, the policy implications for developing technologies common to HDTV and other downstream applications are twofold. First, the United States should be careful not to reduce its support for R&D and sit back and wait to benefit from the work of others. Second, the United States should not cut itself off from the benefit of foreign experience by discriminating against foreign firms in publicly funded research projects. Moreover, if the benefits of U.S. research will spread abroad and vice versa, then from the outset international collaboration in the development of technology should be encouraged.

On the first issue, the importance of a strong commitment to research and the development of new technologies should not be underestimated. U.S. R&D has been the historical bedrock of American competitiveness, and it would be a serious mistake to curtail the nation's commitment during the inevitable budget battles ahead. In particular, private and public research in advanced digital technologies, a core concern raised during the HDTV debate, should be supported. The policy question is how the government can support the development of these technologies without biasing it toward HDTV or consumer electronics applications.

Given the uncertain nature of HDTV's role as a strategic industry for the United States, the policy interest is best defined not by promotion of a consumer electronics product but by the underlying collection of cross-cutting technologies. The American Electronics Association has identified at least seventeen technologies, such as digital display technologies and fiber optics, involved in digital electronic systems that are expected to have applications in at least thirty businesses (tables 6–1 and 6–2). But AEA's HDTV industrial policy conclusion, that the government needs to help create the consumer HDTV industry to maintain the future viability of other U.S. industries, overstates the importance of the consumer HDTV application and understates the complexity of the development and commercialization process.[2]

93

TABLE 6–1

TECHNOLOGIES INVOLVED IN DIGITAL ELECTRONIC SYSTEMS

Video compression	Laser systems
Audio compression	Substrate materials
Digital display	Charge couple devices
Cameras	Flat panels
Videocassette recorders	Data storage and retrieval
Microelectronics	TV production
Fiber optics	Avionics
Opto-electronics	Defense electronics
Cables	

SOURCE: American Electronics Association.

Policy makers should as a result continue to resist proposals that selectively support HDTV as a consumer electronics product or are biased toward its application, such as the industry-led proposals for the ATV Corporation or its stepsister, the Consumer Electronics Capital Corporation, which may draw capital away from more productive uses and increase the general cost of capital. Future proposals to support the development of high-definition technologies should also be resisted unless the projects are both evenhanded in their effect on the various industries and are underfunded by market participants.

"Generic" is the buzz word used to describe research that meets these conditions for federal support. But what is generic research? Practically everyone agrees that more of it is necessary, but few guidelines are available to determine systematically which projects merit public funding. Indeed how the government should support the development of generic technologies remains controversial and is quickly becoming a central issue in the ongoing American industrial policy debate.

In response to questions during a congressional hearing on HDTV, Alan Bromley, science adviser to the president and director of the Office of Science and Technology Policy, characterized generic technologies as those that are necessary for many disciplines, help to support the technological superstructure, and require subsequent R&D to achieve commercial applications.[3]

Translating this definition into particular research projects that merit government support is difficult because of the middle-ground nature of the research, namely, somewhere between basic research and proprietary technology. At the basic end of the research spectrum, the economics are relatively straightforward: ideas are just forming,

TABLE 6–2

APPLICATIONS OF DIGITAL ELECTRONIC SYSTEMS

TV set manufacturing	Software programming
VCR manufacturing	Electronic publishing
Magnetic tape manufacturing	Computer servicing
Video/audio compression	Video test
Cable companies	Retail shopping
Electronic components	CAD/CAM manufacturing
Display manufacturing	Data storage
Circuit manufacturing	Medical imaging
Broadcasting	AI production
Networks	Aircraft avionics
Satellite dish manufacturing	Computer manufacturing
Telephone manufacturing	Graphics software
TV-video programming	Space scanning
Defense electronics manufacturing	Robotics
Laser manufacturing	Weather forecasting
Audio components manufacturing	Navigation

SOURCE: American Electronics Association.

no products are in sight, returns can usually be appropriated by others without permission or compensation, and if investors are not compensated, society will tend to get less basic research than it desires. As a result of relatively simple economics, legitimate basic research projects are relatively easy to identify. In contrast generic research projects fall into a vast gray area. Appropriation problems may provide an economic case for public funding support. But actually picking these projects on the basis of economic criteria is more complicated since research in this middle stage involves translating existing ideas into concepts for products and processes.

In this context a general rule of thumb used by economists seems particularly relevant for federally funded research in high-definition technologies: stay out of the business of making commercialization decisions. The intention should be to support the function of research rather than of a particular firm or industry. Possible research areas, with problems of incomplete appropriability, that may merit government intervention, include projects that are not associated with proprietary product designs and whose results are expected to spread to a variety of other industries. Projects that engage the government in ensuring the commercial viability of a particular product, firm, or industry should be rejected.

Candidates in areas related to HDTV include precommercialization research in high-definition displays, data compression, and advanced digital-switching technologies. Many believe these technologies will have multiple applications in fields such as computers, medicine, and television. To the extent that these technologies are not being adequately funded, the government should serve as a catalyst for their development but should avoid the business of shaping market paths for any of the technologies through measures such as capital subsidies for production.

Approximately $120 million was set aside in the 1990 U.S. federal research budget under the label of advanced imaging technologies.[4] The Defense Advanced Research Projects Agency in particular had selected sixteen company teams by mid-1990 as part of its $30 million high-definition display system program, which consisted of display and processor-related projects[5] (table 6–3). The intent is to "maintain or improve the capability of currently available high definition display technologies while significantly reducing cost" by innovation or maintaining production to realize economies of scale.[6] In the display area DARPA has selected teams to develop projection display technology and plasma, liquid-crystal, and electroluminescence flat panel display technologies. In the processor projects the aim is to increase information-processing capability, and two different modes of transmitting data are being explored. One project is developing advanced coding schemes in which all data are sent at once; a second looks at psychovisual processing theory, whereby only the data needed by the human eye are sent and the human brain is basically fooled.

Some of these projects that focus on the delivery of preprototype equipment raise serious questions about how far the government should venture toward the marketplace to promote economic objectives. If the intention of the project is to upgrade existing equipment to improve the commercial viability of selected firms, then the government has strayed too far into the commercialization end of the development process. Some of the Sematech projects conducted under the equipment improvement program, for example, seem to fit this description.[7] If the objective is, however, to fill a gap left by the market and to support the research process open to a wide range of industrial interests, then government support conforms more to the notion of generic research and is thus more appropriate.

At a time when total funding levels are the subject of congressional budget battles, increased funding for dual-use technology projects through the Defense Department may also limit the funding for military-specific technologies and applications that are most likely *not* to be funded by the private sector. In this environment the DARPA high-definition display projects need to be viewed as a series of trade-offs. The nation should not necessarily accept whatever the market produces to satisfy national security objectives; instead the Defense Department should focus on those critical areas that it determines cannot be supplied with sufficient reliability by the market and accept that the department may not be the appropriate place to support the development of dual-use technologies whose primary applications are expected to occur in the fast-paced, cost-sensitive commercial sector.[8]

HDTV has also raised the issue of performance requirements, such as local content as a precondition for access of foreign firms to public-private research programs. This consideration, however, has taken the U.S. debate on an untimely digression away from the more fundamental economic concern of how to increase and to maintain an attractive U.S. environment for investment and production over the long term. Rather than trying to get the basics right to maintain an attractive investment environment, artificial barriers have been suggested as a crude attempt to lock in the investment of footloose firms. Moreover it would be difficult, if not impossible, to stipulate that state-of-the-art research be done here. Similarly, if a link is required between research and final production, monitoring it could easily become a procedural nightmare especially as firms continue to weave an international web of research and production activities.

Finally, the importance of federal research support for generic technologies should not be overstated: developing the technologies is only one element in the process of improving the performance of the nation's resources. The extent and speed of adoption of technological advances play a central role. Thus obstacles to diffusion and adoption are at least as important as the initial development of these technologies.[9] Structural and regulatory factors may unnecessarily impede the adoption and adaptation of some of these technologies in the United States. Standards that encourage interoperability will likely ease the diffusion process. It may also be necessary to reevaluate how the interests of innovators and of consumers are balanced so that greater diffusion is encouraged and advances are incorporated into general practice in the economy.[10]

A final note is needed on the issue of production location. The effects of particular industries on the national welfare are still being explored by economists and policy makers. More important, if the location of manufacturing in the United States over the long term is the root concern, then it should be addressed directly. HDTV does not warrant special consideration. Instead of policies targeting strategic industries, policies more uniform in orientation should be considered, such as those focused directly on building and maintaining strength

TABLE 6–3

THE DARPA HIGH-DEFINITION DISPLAY PROGRAM

Firm	Project
Display Technology[a]	
Projectavision	Active-matrix LCD technology
NewCo Inc. (Nitcor)	Laser projection display technology
Xerox	Thin-film transistor-driven liquid crystal display, flat panel, and projection system
Texas Instruments	Projection display technology (TI as team leader, Sarnoff as a subcontractor)
Photonics	Plasma flat panel display
Magnascreen Corp.	Technology for large, full-color active-matrix flat panel displays for military applications
Planar Systems	Technology to permit manufacturing of full-color flat panel displays based on electroluminescent technology
Zenith Electronics	Interchangeable flat tension mask technology for advanced CRT displays
Norden Systems	Low-resistivity electrodes for large-area electroluminescent displays
MRS Technology	Large-area photo lithography tool development for active-matrix displays
Ovonic Imaging Systems (Optical Imaging Systems)	Manufacturing technology for large-area LCDs
Microelectronics and Computer Technology Corp. (MCC)	Field emission flat panel displays
Tektronix	Plasma-addressing liquid crystal technology
Display Processor Development[b]	
Sarnoff, Sun, Texas Instruments	High-definition video workstation (Sarnoff as team leader, providing processor and software; Sun providing workstation platform and architecture)
	Multimedia workstation including high-resolution display and interactive video input (Sun and Sarnoff working together)

TABLE 6—3 *(Continued)*

Firm	Project
	Custom semiconductors based on computer workstation research liaison with DARPA and rest of industry (TI)
Adams Russel Electronics	Advanced compression MIT technology based on MIT's previous research
Qualcomm	Alternative to Russel/MIT compression technology based on cutting-edge technology in digital communications
Display Processor Design[b]	
Sarnoff	Advanced computing technology in support of display processor design, based on the assembly of system at National Institute of Standards and Technology (NIST) for high-definition display research; the Princeton engine as the base permitting rapid (real-time) evaluation of new concepts in signal processing, including compression of signals for moving images, for an array of applications

a. Projects announced June 1989 and August 1990.
b. Projects announced October 1989.
SOURCES: Defense Department news releases dated June 13, 1989, October 26, 1989, and August 14, 1990; *Air Force Magazine*, April 1990.

in domestic factors of production: a sound education system, kindergarten through grade 12; a technically trained work force and a sophisticated transportation and communications infrastructure.

International Policies

At the international level a major policy challenge illustrated by the HDTV debate can be summarized in one word: access. If it is limited, then the ability of the United States and other nations to appropriate gains from scale economies or innovations developed here or elsewhere will be also limited. More important, perceptions of unequal access have sparked fears of dependence on foreign products and have fueled demands for protective industrial policies. The factors behind access problems are multiple, and the answers are uncertain. But inward-looking policies are clearly part of the problem, not the solution.

Consider, for example, the European response to the competition of Japanese and U.S. firms. The Europeans tried to build independent domestic capabilities in advanced technologies through publicly supported programs such as Esprit, Eureka, and Jessi. They are, however, discovering the hard way that global competition, international alliances, and mobile multinationals have combined to limit the cost-effectiveness of this approach. Their programs, intended to strengthen domestic firms and to counter Japan's strengths, are coming up short. Among the many problems they cannot even determine which firms are European. The debate is intense, sparked in part by two recent events: Fujitsu's recent purchase of ICL, Europe's most profitable computer company, and ICL's subsequent expulsion from three of the Jessi projects.

Meanwhile, on this side of the Atlantic, Sematech members are having second thoughts on how to improve their competitive position: at least half of the chip makers in Sematech have joined forces with Japanese firms through a variety of research, production, and marketing arrangements to develop advanced chip technologies and products. Yet the consortium explicitly excluded the Japanese because it was intended to ensure an independent domestic base in semiconductor technology. Together the debate over ICL in Europe and the actions of American chip makers underscore the fact that looking inward has resolved little but may cost both the individual firms and nation a great deal.[11]

Therefore, if the benefits of U.S. research in advanced imaging technologies and high-definition systems are going to spread abroad, international collaboration in the development of the technologies should be encouraged rather than impeded. Whatever research projects are selected for federal support should not be biased against participation of foreign-owned firms. There is no reason for the United States to start from scratch and to exclude companies such as Philips and Thomson that have already invested heavily in HDTV-related areas. Moreover, if the general objective is to increase the diffusion of high-technology innovations within the U.S. economy, then severing the United States from the growing web of alliances between these firms and other international research efforts is counterproductive: The "not invented here" syndrome is out of place in a time of global integration and technology-driven international investment strategies.

On the flip side the United States should continue to push for access of U.S. firms to international HDTV-related research activities.

The Japanese broadcast agency NHK has offered to license the technology behind the MUSE or Hi-Vision system to foreign firms. The Electronics Industry Association of Japan and the U.S. Semiconductor Industry Association are promoting international business forums to encourage further the participation of foreign firms in HDTV-related research and production.

Proposals for international collaboration should be further encouraged in the development of advanced digital technologies that underlie products such as HDTV and telecomputers. The alliances can serve a variety of useful purposes such as pooling development risk, extending the range of experience, diversifying the emerging base of suppliers, improving the ability of U.S. firms to learn from the successes and failures of other countries, and improving market access.[12] Video-processing circuitry and digital compression technologies are particularly fertile areas for international research alliances because of the high R&D costs, recent advances in digital video transmission and U.S. strengths in designing powerful processing chips, and Japanese strengths in consumer electronics.[13]

Finally, and arguably most important, the HDTV debate has highlighted the need for stronger multilateral mechanisms to address differences in industrial practices and market structures that affect trade and market access. One concern in the HDTV debate, for example, has been that the access of U.S.-based semiconductor suppliers to the emerging HDTV and other product markets in Japan would be limited. Another concern has been that the delivery of advanced components would be delayed, damaging the competitiveness of U.S. producers or forcing them to shift their operations abroad.

These concerns are not unique to any one industry and therefore cannot be effectively addressed with industry-specific policies. They nonetheless have demonstrated the potential to strain the international trading system by stimulating anxious calls for defensive strategic industrial policies. The dangers of ignoring these concerns are real. In particular, the competition between nations to build independent technological capabilities and capacity in industries believed to have strategic value (steel, autos, and more recently semiconductors) have disrupted trade, have incited disputes over issues like local content, and threaten to fragment markets further.

Firms may as a result find it difficult to adjust to the rigorous demands of global competition if nationalistic inward-looking programs continue to multiply. The pressures from such programs are already apparent. Examples include the problems of surplus production capacity stimulated by promotion (protection) plans in the steel and auto industries. In addition major firms in the semiconductor

101

industry have been forced to set up additional production facilities in Europe to avoid the sting of local-content restrictions and antidumping duties. Rather than encourage this ill-fated path, the United States should discourage it and continue unwavering support for the development of a stronger multilateral system capable of releasing some of the steam now powering industrial policy demands at home and abroad.

Furthermore the United States could not pick a worse time for boosting the case of industrial policy. The United States has long criticized the Europeans for their heavy industrial subsidies as well as their protection of national champions. In recent years the EC has made important progress toward reforming the subsidy system and is in the midst of a passionate internal debate over the future of industrial policy prescriptions. U.S. moves toward a program of picking and promoting strategic industries would undoubtedly—to the detriment of national long-term interests—undercut the position of those pushing for additional reforms and bolster the case of their opponents.[14]

Additionally the industrial policy approach seems to have enhanced, rather than curbed, national fears of rising interdependence. The prisoner's dilemma, often discussed in texts on game theory and psychology, is similar in concept. Two partners in crime have been arrested and kept in separate cells. The fate of each is dependent on the other, but they cannot communicate to create a mutually beneficial story. In isolation both prisoners have an incentive to label the other as the instigator and to lessen their own punishment. If both prisoners do this, however, both will get harsher sentences than they would have received if they had coordinated their stories and stayed silent. The policy lesson from the prisoner's dilemma is that it pays to communicate and to recognize that actions are interdependent. Unlike the prisoners the United States and other countries have this option but have failed to construct the means with which it can be effectively exercised with respect to advanced technologies and their commercialization.

To move closer to a system in which interdependence is viewed more as an opportunity and less as a risk to be avoided at all costs, international policy efforts should focus on extending the coverage of multilateral mechanisms to those policies that have traditionally been considered domestic but distort trade and international investment flows.[15] The examination of previously tolerated domestic investment subsidies should be further encouraged. The United States should also seek international agreement on acceptable responses to offset injury caused by industrial policies. Beyond GATT, reform efforts should continue by creating a credible multilateral mechanism to discuss differences in national competition policies and practices.[16]

As a result of differences in political-economic structures, customs, and historical circumstances, international negotiations in all of these areas will be difficult and tortuous. Yet the cost of delaying the design of these forums is likely to escalate if the credibility of the multilateral systems of rules-based trade continues to be undercut by strategic industrial policies.[17] Moreover the postwar period has demonstrated that no other path exists with the same benefits for predictability and global prosperity as a strong, multilateral trading system. Domestic policy should in the meantime focus on encouraging the development and widespread diffusion of advanced technologies to the U.S. economy by supporting the research technologies expected to have extensive applications and promoting an attractive investment and production environment for all firms regardless of nationality.

7
A Guide for Future Debates

In keeping with the Bush administration's ban on industrial policy, those in favor of more activist technology policies in the United States frequently argue these days that picking and promoting winners is not part of the agenda. They suggest the United States move toward improving its economic future by getting the priorities right and listing the technologies critical for future economic growth. As noted, more ambitious and industry-specific plans are around the corner. In particular, a bill introduced by Congressman Norman Y. Mineta (D, Calif.) in the spring of 1990 calls for the secretary of commerce to devise a list of strategic industries that "will provide the bulk of economic opportunity and economic growth." A problem with making lists, however, is deciding whom to include and to exclude, which sounds suspiciously like moving further down the ill-fated path of picking winners and losers.

Therefore, although the calls for a national HDTV plan are no longer front-page news or the popular subject of editorials, the pressure persists for the government to tilt toward industrial policy to support the commercial development of advanced technologies in the United States. The HDTV debate has confirmed the misgivings of skeptics and demonstrated the weaknesses of the activist case. It would serve the nation well to keep these lessons in mind as public officials and their advisors turn to the advice of the various list-making groups that seek to influence policy. Four in particular merit attention.

Do Not Confuse Spillovers with Strategic Importance. If technological and linkage spillovers alone signaled the strategic importance of an industry, then virtually all industries would be strategic. But this conclusion does not follow from the premise that strategic industries are more important than other industries.

Further, as a new model for developing all other key industries in the future, HDTV has ironically demonstrated why skeptics are skeptical about whether a policy of strategic industrial targeting could be welfare-enhancing. Despite a decade of research into the policy

104

implications of these industries, policy makers have demonstrated that they are effectively no better in knowing how to pick or to support strategic industries. In short it is a policy gamble. And if the recent HDTV debate is any guide on what can be expected by instituting a policy of strategic industry promotion, the results will not be worth the risk.

Be Wary of "Generic" as well as "Strategic" Labels. The digital revolution is creating new opportunities for a variety of industries to benefit from technological advances in other industries. This situation dramatically complicates the issue of strategic industry promotion because in a world of two-way spillovers difficult choices must be made about which industry or subindustry is more important for capturing the returns associated with the development of the under-lying technologies and components. This is a vexing task if, as with HDTV, the underlying technologies and components have multiple applications that may precede the emergence of the target industry. Policy makers should thus be wary of proposals that focus on only one industry's application of the underlying technologies. Such an approach tends to oversimplify the process of technological change and the problems of successful commercialization of new technologies in the United States.

The HDTV industrial policy proposals, industry-led or otherwise, further confirm that support for a strategic industry candidate would involve much more than research support for advanced technologies: instead at issue is a long-term stake in a particular industry's domestic development. Additionally the economic case for government support of generic research is fundamentally different from the one for strategic industries. Support for generic research is by definition broad in orientation while the other is selective. The former does not try to pick a development path, but it is the objective of the latter. In cases such as the HDTV industry, one defines a limit for government support while it is effectively left open in the strategic industry case because of the decision to take a long-term national stake in the domestic development of the industry.

Do Not Compromise: Industrial Policy Is Not a Solution. The issue of how government should promote national participation in technolog-ically promising industries will increase in importance as nations continue to compete for the investment of footloose firms in the 1990s.

The HDTV debate demonstrated the potential for industrial targeting to miss the mark and to encourage moves away from a more open world trading system toward one in which nations try to create indigenous technologies and to nurture particular industries. But policy makers should recognize that few technologies are national in scope and whatever edge industrial policy may have had in the past is being eroded by the globalization of competition and the mobility of multinational corporations.

Conversely the costs of such an approach may be considerable, especially to the international trading system. Interdependence and risk are inherent characteristics of the evolving global economy. Industrial policies cannot change this. Further, by matching targeting with countertargeting for HDTV or future strategic industry candidates, industrial policy reinforces the dangerous precedent of the semiconductor industry, which could lead to rising trade tensions and an unraveling of the international trading system. The tendency for each nation to make strategic industry decisions alone as if each were independent without a long-term interest in the larger trading system must be checked, not encouraged. A policy of strategic industry promotion on the part of the United States would in this respect be part of the problem, not the solution.

Get the Basics Right. Given that competition between nations for the investment dollars of transnational corporations seems likely to continue to increase, the question is not so much whether but where investment in advanced technologies will occur. The United States had the comfort of political stability, unequaled research capabilities, the world's largest consumer market, and one of the world's largest and most efficient capital markets to attract investors. The emergence of a single European market will challenge this comfortable position and will intensify the competition for investment. To meet this challenge effectively, the United States will be increasingly pressed to maintain an attractive investment environment and to make efficient use of the country's considerable resources. Being a better industry partner in this respect does not require industrial policy, but it does require investing in a skilled work force and supporting research that would otherwise not be undertaken.

To conclude, exploring HDTV as a strategic industry candidate has raised more reasons to suspect than to support the case for industrial policy. The administration and Congress should continue to pick their way carefully through the inevitable array of high-technology policy challenges that have recently revolved around HDTV and that are likely to revolve around other candidates in the 1990s. Rapid changes in technology and the emergence of sophisticated economic peers do challenge the United States to respond. But this response should be grounded in reasoned analysis of what is known, what is politically feasible, and what is likely to be constructive over the long term. Changes in the global economy have complicated the definition of national technological capabilities and have done little to reduce the difficulty of picking today which industry may become strategic for the United States. More than anything else, the lengthy two-year policy debate over HDTV as a strategic industry has illustrated this lesson.

Appendix

The development of an HDTV system depends on the interrelated standards for transmission and production of information. The production standards govern the creation of HDTV programs while the transmission standards govern the delivery of these programs to HDTV users such as household consumers. In an analog tranmission system, a higher standard in production is necessary to cover the losses in quality during the broadcasting process. A digital system circumvents this problem. Transmission standards will affect potential market participants as the producer equipment makers are expected to build to what the FCC selects. How the HDTV signal is broadcast thus affects how it is produced and displayed. Special coding is needed to compress the HDTV signal and to minimize the required space in the radio spectrum. Designing sophisticated, reliable, and affordable compression techniques is one of the primary challenges before the emerging HDTV industry. The lure of the digital marketplace and the pressure of rivals have combined to intensify competition in the domestic television industry.

As of mid-1990 six systems were scheduled for testing by the privately funded Advanced Television Test Center and then for FCC consideration. The FCC has announced two major objectives for this process. In September 1988 the commission announced that transmitted signals must be compatible with existing NTSC sets and that no additional spectrum would be allocated for television. This was followed by an announcement in March 1990 of the FCC's preference for a simulcast system. Under this approach two distinct signals would be sent: a HDTV signal and a conventional TV signal. Sharper images would be available to consumers who owned an HDTV receiver, while other consumers would continue to view the standard-quality images on their existing sets. Eventually the broadcast of the conventional NTSC signal would be phased out, and all consumers would then need a HDTV receiver to watch broadcast television.

To create a sharper picture, HDTV signals need to increase dramatically the information that conventional NTSC signals incorporate. Accordingly the delivery of this signal poses a considerable technical challenge for all the proposed transmission systems. A 30-megahertz HDTV signal must be squeezed into a 6-megahertz bandwidth (a single TV channel) allocated by the FCC. Some engineers have likened this signal compression process to squeezing an elephant into a bathtub.

A digital system has been called the wave of the future as it represents the threshold to a new age in entertainment and information for television, where sets will have the capability to receive, to manipulate, and to store video information. Before 1990 no one had demonstrated, however, the compression of a full HDTV signal to a single channel or end-to-end digital transmission. Digital TV was therefore not expected by technology experts until the twenty-first century.

Digital HDTV and interactive television may, however, be coming sooner than expected. A turning point was reached in mid-1990 when General Instruments stunned its rivals at home and abroad with its proposal to the FCC for an all-digital system (Digicipher). This system would transmit programming as a series of ones and zeros rather than the analog waves of conventional television. After this announcement, competition between the research labs intensified; within six months two of the other contenders had made rapid progress. Zenith with AT&T as well as the David Sarnoff Research consortium with Philips, Thomson, and NBC had replaced their initial proposals with fully digital systems for testing. Given these developments and the prospect of adopting a fully digital simulcast system, the FCC has also announced that an intermediate step such as EDTV may be desirable to provide the time to finish the research and eliminate defects of a digital system.

If successful in this process, the United States is likely to leap-frog Japan. Indeed Japan has largely missed out on most of the recent advances in the fully digital systems because of its early decision to base its system on conventional analog technologies. Instead of tackling the video data compression problem, Japanese planners decided to choose a less elegant solution to the bandwidth dilemma, namely, expanding bandwidth and relying on satellites for signal delivery. This path for HDTV signal transmission is currently incompatible with the terrestrial or over-the-air broadcasts that control more than 50 percent of the U.S. television audience. This approach is also less sophisticated, relies on conventional analog transmission technologies, and as a result may dilute technological spillover effects for other industries. Although satellite transmission of HDTV programming is limited in Japan, wider coverage is expected sometime in

1992. Japan is also now considering EDTV as an intermediate step for use by broadcasters and satellite operators.

To the extent that the signal compression problems are not resolved before testing concludes, other technical and political problems regarding spectrum allocation must be overcome. The political problems involve finding channels for HDTV signals in the current allocation of spectrum. In theory many "new" channels could be created by making use of the so-called taboo channels that separate conventional television channels. The technical challenge involves applying new technology to reduce the interference between signals and to permit effective use of the taboo channels.

A third challenge is the cost of conversion. For a studio this involves shifting from the production of TV programs based on the conventional format to the production of HDTV programs. NBC has estimated this cost for its network conversion at $500 million. For a TV viewer, this means the purchase of a new TV set. Such sets were introduced at $34,000 in Japan. Although the price of U.S. sets is expected to be considerably lower, they will command a premium price over conventional sets if a fully digital transmission system is adopted.

For a long-term solution for HDTV signal transmission, industry experts have focused on broader band media. In the United States attention has focused on fiber optics because of its high-bandwidth advantages and connection with other future information-processing industries. In Europe and Japan direct satellite broadcast systems have already been chosen.

Notes

INTRODUCTION

1. "An Open Letter to President Bush and Congress," dated October 9, 1989, distributed by Rebuild America, Washington, D.C.

2. Council on Competitiveness, *Gaining New Ground: Technology Priorities for America's Future* (Washington, D.C.: CoC, March 1990).

3. Anne O. Krueger, "Free Trade Is the Best Policy," in Robert Lawrence and Charles Schultze, eds., *An American Trade Strategy: Options for the 1990s* (Washington, D.C.: Brookings Institution, 1990), p. 84.

4. The nature of the HDTV debate requires analysis of an array of technical, economic, and political issues. Accepting the benefits from specialization of labor, this study has relied on the analysis of engineers and other technology experts for insight on the technical questions such as the maturity of some of the technologies involved and the potential for interindustry technological spillovers and the form they may take as the development process continues.

5. Under certain conditions, by favoring domestic firms in the industry, government may alter the behavior of foreign firms so that a larger share of rent is shifted to domestic firms and national income may increase as a result. Analysts have, however, criticized this proposition as too assumption dependent and information intensive. In addition, if another country retaliates, the payoff matrix for the nation is fundamentally altered and the rationale for intervention is undercut.

6. Strategic industries have also been defined in terms of their importance for military strategies and objectives. Neither the military nor the rent-shifting arguments for strategic industry promotion are explored in this study. Instead the study focuses on the interindustry spillover argument frequently advanced in cases such as HDTV to justify preferential industry treatment.

7. Policy makers, for example, would need to have the foresight to predict which industries will be essential to the national welfare, the extent to which the private sector is underinvesting in their development, and how support for their development could be structured so that average productivity is enhanced rather than curtailed.

8. Fred Branfman, Rebuild America, in U.S. Congress, House, Subcommittee on Telecommunications and Finance of the Committee on Energy and

Commerce, *Hearings on High Definition Television*, 101st Congress, 1st session, pp. 193–204.

9. Representative Don Ritter, cochairman of the House HDTV Caucus, "To Miss Out on High Definition TV Is to Miss Out on the 21st Century," *Roll Call*, Competitiveness Policy Briefing, July 10–16, 1989, pp. 16–17.

CHAPTER 1: DIMINISHED GIANT SYNDROME

1. For a recent study see National Academy of Engineering, *Prospering in a Global Economy* (Washington, D.C.: National Academy Press, 1991).

2. Based on the U.S. Department of Commerce DOC-3 definition of high-tech industries, which includes all products within a high-tech industry group regardless of the level of technology embodied in the product. National Science Board, *Science Indicators 1989* (Washington, D.C.: U.S. Government Printing Office, 1989), appendix tables 7–9. See also J. Hayden Boyd, "Manufacturing Growth and the Trade Deficit: Does Trade Cause the Decline of Industries?" State Department Working Paper (November 1988).

3. U.S. Department of Commerce, International Trade Administration, *United States Trade Performance in 1988* (Washington, D.C.: U.S. Government Printing Office, 1989), p. 24. Trade in technology-intensive and capital-intensive industries is illustrated by the office equipment industry, the second largest export industry and the third largest import industry in 1988. There is some doubt about the degree to which the phenomenon of intra-industry trade can be explained by methods of aggregating the data. Sound economic reasons remain, however, why trade in similar products is desirable for the sake of scale economies and product differentiation. See, for example, Wilfred Ethier, *Modern International Economics* (W. W. Norton, 1983), pp. 42–47.

4. National Academy of Engineering *Prospering in a Global Economy*, p. 25.

5. See William J. Baumol, "Productivity Growth, Convergence, and Welfare: What the Long-run Data Show," *American Economic Review*, vol. 76, no. 5, pp. 1072–. See also Charles Hulten, "From Productivity Slowdown to New Investment Opportunities: A Literature Review" (Paper prepared for Conference on International Capital Flows, Kiel Institute of World Economy, June 1990).

6. Robert Litan, Robert Lawrence and Charles Schultze, eds., *American Living Standards* (Washington, D.C.: Brookings Institution, 1989), pp. 24–25, 187.

7. Jagdish N. Bhagwati, "U.S. Trade Policy Today" (Paper presented at the Columbia University Conference on U.S. Trade Policy, September 8, 1989), p. 29; National Science Board, *Science Indicators 1989*, table 7–8. There is considerable debate about the magnitude and meaning of the U.S. decline in world market shares in the post–World War II period. For a useful analysis of mistakes often made in interpreting the data, see Sven Arndt and Lawrence Bouton, *Competitiveness: The United States in World Trade* (Washington, D.C.: American Enterprise Institute, 1987), chap. 3.

8. Bhagwati, "U.S. Trade Policy Today," pp. 27–30.

9. Ibid. See also Michael Porter, *The Competitive Advantage of Nations* (New York: Free Press, 1990); and Herbert Stein, "The U.S. Economy: A Visitor's Guide," *The American Enterprise*, July-August 1990.

10. U.S. Congress, House, Subcommittee on Telecommunications and Finance of the Committee on Energy and Commerce, *Hearings on High Definition Television*, 101st Congress, 1st session, March 1989. See also Lester Thurow, Fred Branfman, George Lodge, and Ed Miller, "Fiddling While U.S. Industry Burns" (White paper, Rebuild America, Washington, D.C., 1990).

11. Robert J. Samuelson, "Marshall's Mixed Legacy," *Washington Post*, June 17, 1987, p. F-1.

12. Different measures of high-tech trade yield, however, dramatically different interpretations of U.S. performance in the 1980s. The data most frequently cited in the press are based on the International Trade Administration's definition of "high tech" described in note 2 that suggests that the U.S. high-tech trade balance turned negative in 1986 and then slowly improved in subsequent years. The Bureau of the Census offers a more detailed and accurate picture of the trade balance in high-tech products based on a definition of advanced technology products. The ATP measure remained positive throughout the late 1980s. See Thomas Abbott and Robert McGuckin, "Measuring the Trade Balance in Advanced Technology Products" (Center for Economic Studies, Bureau of the Census, Washington, D.C., January 1989).

13. For a useful discussion of the relative importance of macroeconomic and microeconomic factors for U.S. competitiveness in the 1980s, see Arndt and Bouton, *Competitiveness: The United States in the World Economy*.

14. House, Subcommittee on Telecommunications and Finance of the Committee on Energy and Commerce, *Hearings on High Definition Television*, particularly the comments of Fred Branfman, director of Rebuild America, and Representative Ritter, cochairman of the House HDTV Caucus, p. 182. See also Paul Bluestein and Evelyn Richards, "U.S. Weighs Industrial Policy Shift," *Washington Post*, May 7, 1989.

CHAPTER 2: THE STRATEGIC INDUSTRY DEBATE

1. For an overview and history of the debate, see Claude Barfield and William A. Schambra, eds., *The Politics of Industrial Policy* (Washington, D.C.: American Enterprise Institute, 1986).

2. Well-known U.S. exceptions to the no-targeting rule include support for agriculture and industries affected by defense interests. How successful industrial policy has been in civilian industries is still a matter of dispute, while others question the wisdom of trying to shift the focus of industrial policy from defense industries to commercial markets. Among the many references on the issues, see Richard Corrigan in "Smokestacks and Silicon— The Debate over U.S. Industrial Policy Continues," in *Smokestacks and Silicon: Regaining America's Edge* (Washington, D.C.: National Journal, 1984), pp. 6–10. On the problems of mixing military procurement and commercial objectives, see Richard R. Nelson, *High-Technology Policies: A Five-Nation Comparison* (Washington, D.C.: American Enterprise Institute, 1984), especially

pp. 72–77. See also Paul Geroski, "Brander's 'Shaping Comparative Advantage': Some Comments," in Richard Lipsey and Wendy Dobson, eds., *Shaping Comparative Advantage* (Toronto: C. D. Howe Institute, 1987). For a useful comparison of the American and Japanese approaches toward industrial policy, see Daniel Okimoto, *Between MITI and the Market: Japanese Industrial Policy for High Technology* (Stanford, Calif.: Stanford University Press, 1989), especially pp. 49–54.

3. Determining whether an industry is high-tech is, however, difficult because some of the traditional industries use advanced technologies, materials, and equipment in the production process.

4. For examples of this argument, see Jeff Faux, "Whatever Term You Use, 'Industrial Policy' Looks Like an Idea Whose Time Has Come," *Roll Call*, Competitiveness Policy Briefing, no. 4, July 1989, pp. 22–23; testimony of Fred Branfman and Pat Choate on national champion programs abroad and implications for the United States, in U.S. Congress, House, Subcommittee on Telecommunications and Finance of the Committee on Energy and Commerce, *Hearings on High Definition Television*, 101st Congress, 1st session, March 1989; Rebuild America, *Consortia and Capital: Industry-led Policy in the 1990s* (Report prepared for the "Wake-Up America!" conference, May 1989, Washington, D.C.).

5. For a good summary of the different definitions of industrial policy, see Richard G. Lipsey, "Report on the Workshop," in Lipsey and Dobson, eds., *Shaping Comparative Advantage*, pp. 132–38.

6. United States, International Trade Commission, *Foreign Industrial Targeting and Its Effects on U.S. Industries, Phase 1: Japan* (Washington, D.C.: Government Printing Office, USITC 1427, October 1983), p. 17.

7. For critical review of the arrangement, see Kenneth Flamm, "Making New Rules," *Brookings Review*, vol. 9, no. 2 (Spring 1991), pp. 22–29.

8. Adjustment or transition costs have also been cited as a potential rationale for intervention as distinct from market failure. But disagreement exists in the economics profession over treatment of these costs. See, for example, James A. Brander, "Shaping Comparative Advantage: Trade Policy, Industrial Policy, and Economic Performance," and Paul A. Geroski, "Some Comments," in Lipsey and Dobson, *Shaping Comparative Advantage*, pp. 59–61.

9. Various factors, rather than a single theory, are used these days to explain comparative advantage or a nation's pattern of trade. Differences in climate or in natural resources or factor costs are the most familiar explanations. More recent additions involve the scale of manufacturing operations as when the high-volume output is advantageous so that a single manufacturer can service several markets.

10. Perfect competition assumes that price is an accurate indicator of social costs and benefits and that market participants are small and autonomous (that is, they lack the power to determine the market price). Trade opportunities are assumed to be independent of trade policy. Well-known exceptions to this mold include the optimal tariff or large-country case of intervention to exploit market power to improve the nation's terms of trade.

116

11. Many ideas in the new trade literature are still evolving, and the list of references is long and growing. For the classic reference work, see Paul Krugman, *Strategic Trade Policy and the New International Economics* (Cambridge, Mass.: MIT Press, 1987). For a detailed technical discussion, see Elhanan Helpman and Paul Krugman, *Trade Policy and Market Structure* (Cambridge, Mass.: MIT Press, 1989). For summary pieces of the theory and the policy debate, see Krugman, "Strategic Sectors and International Competition," in Robert Stern, ed., *U.S. Trade Policies in a Changing World Economy* (Cambridge, Mass.: MIT Press, 1987), pp. 207–39, and "Is Free Trade Passé?" *Economic Perspective*, vol. 1, no. 2 (Fall 1987), pp. 131–44. For a particularly skeptical review, see Jagdish Bhagwati, "Is Free Trade Passé after All?" *Weltwirtschaftliches Archiv*, vol. 1 (1989), pp. 17–44.

12. An 1987 report from the Organization for Economic Cooperation and Development found that the science-based, scale-intensive, and differentiated commodities and services share of total trade and production for fourteen large OECD countries had increased in importance over the past two decades while the resource and labor-intensive commodity share had declined; see *Structural Adjustment and Economic Performance* (Paris: OECD, 1987). For a basic description of the link between the intra-industry expansion of trade and scale economies among industrialized countries, see Wilfred Ethier, *Modern International Economics* (New York: W. W. Norton, 1983), pp. 42–47.

13. History in this context includes both what might be described as an accident and as more deliberate attempts to alter conditions that attract firms to locate within a particular country or region over the long term.

14. These mechanisms may include the advantages that a firm may gain from being the first to market with a successful product, which can translate into an opportunity for greater investment and a jump on its rivals in developing and marketing the next generation of the product, thus continuing the cycle. Or positive feedback may take the form of R&D in one industry increasing the marginal return to research in another domestic industry. See Joseph E. Stiglitz, "Perspectives on Economic Development: Markets, Market Failures, and Development," *American Economic Review*, vol. 79 (May 1989), pp. 198–200, and Giovanni Dosi, Laura D'Andrea Tyson, and John Zysman, "Trade, Technologies, and Development: A Framework for Discussing Japan," in Johnson, Tyson, and Zysman, eds., *Politics and Productivity*. See also Paul Krugman, *Rethinking International Trade* (Cambridge, Mass.: MIT Press, 1990), chap. 7.

15. Bruce R. Scott, "Competitiveness: Self-Help for a Worsening Problem," *Harvard Business Review*, July-August 1989, p. 118.

16. Some argue that the current distribution of investment resources, rather than the magnitude of the investment and savings pool as determined largely by macroeconomic factors, can make a difference. Some of the ideas behind this notion of strategic industries have a long history and are frequently linked to Joseph Schumpeter and his writings on business cycles and economic progress. A number of macroeconomists, however, remain skeptical about the relative importance of these so-called strategic industries for the average

rate of growth in the economy. The relationship between a particular industry and the average rate of productivity and economic growth in the economy is a link that is still being explored in the macroeconomic and microeconomic literatures. References on the link between Schumpeter and the current policy debate include Tyson, Dosi, Zysman, "Trade, Technologies, and Development," and Nelson, *High-Technology Policies: A Five-Nation Comparison*.

17. Robert Litan, Robert Lawrence, and Charles Schultze, *American Living Standards* (Washington, D.C.: Brookings Institution, 1990), p. 62. See also Lawrence, "Innovation and Trade: Meeting the Foreign Challenge," *Review of Economics and Statistics*, vol. 60 (February 1978), pp. 162–63.

18. The structural factors that have contributed to the decline in U.S. terms of trade and world trade shares include a decline in the U.S. world share of capital and skilled labor and the emergence of sophisticated technology peers in Europe and Japan. See Paul R. Krugman, "The U.S. Response to Foreign Industrial Targeting," *Brookings Papers on Economic Activity*, no. 1 (1984), pp. 85–87.

19. For an example of the debate over dumping in the U.S. television industry, see Jeffrey Hart, "The Consumer Electronics Industry in the United States: Its Decline and Future Revival," in Office of Technology Assessment, *The U.S. Electronics Industry Complex* (Report to the U.S. Congress, October 1988).

20. The process of technological development is often associated with a number of market imperfections and dynamic effects. These include problems of appropriation that may arise when an advance is readily copied by rivals. As a result the return cannot be fully captured by the innovator, which may deter investment in socially beneficial activities. Technological development is also frequently associated with increasing returns in the form of learning or dynamic scale economies. A learning economy refers to the concept that experience in production or cumulative output determines current productivity. References include Krugman, *Rethinking International Trade*, and Paul Romer, "Increasing Returns and Long Run Growth," *Journal of Political Economy*, vol. 96 (October 1986), pp. 1002–22.

21. For a recent example of the importance assigned to technology-intensive industries, see Council on Competitiveness, *Gaining New Ground: Technology Priorities for America's Future* (Washington, D.C.: Competitiveness, March 1990). The council suggested in particular that U.S. prosperity depends on telecommunications, drugs, aerospace, construction, electronic components, chemicals, computers, machine tools, and motor vehicles.

22. The FSX debate within the administration and Congress over the transfer of technology to Japan in a joint venture to build a tactical fighter revealed a fundamental fear that the transfer would boost Japanese efforts to build a commercial aerospace business and undermine the position of American producers. See David E. Sanger, "The Technology That the U.S. Doesn't Want Japan to Have," *New York Times*, April 2, 1989. For a critical review of the issues, see David Mowery, "Effects of the FSX Agreement on the Japanese Commercial Aircraft Industry" (Report prepared for the U.S. Congress, Senate Committee on Foreign Relations, 101st Congress, 1st session, May 8, 1989).

23. For references on strategic industries and technological progress as an endogenous variable, see Dosi, Tyson, and Zysman, "Trade, Technologies, and Development," p. 22. See also Krugman, *Rethinking International Trade*, pp. 165–67.

24. Paul Krugman has been a leader in developing the new trade literature. A particularly useful article, written in 1983, retains its relevancy today on the pitfalls of conditions frequently advanced to identify strategic industries. Krugman, "Targeted Industrial Policies: Theory and Evidence, in *Industrial Change and Public Policy* (Kansas City, Mo.: Federal Reserve Bank of Kansas City, August 1983), pp. 123–34.

25. See also Schultze, "Industrial Policy: A Dissent."

26. "New" is a term that has been applied to the trade literature on strategic industries, but a number of the ideas are not so new. Rather than the ideas, the tools that now permit ideas on scale economies to be modeled and tested are new. For other reservations about how new the strategic industry literature is see Jagdish Bhagwati, "Is Free Trade Passé?"; Gottfried Haberler, "Strategic Trade Policy and the New International Economics: A Critical Analysis," in Ronald W. Jones and Anne O. Krueger, eds., *The Political Economy of International Trade* (Cambridge, Mass.: Basil Blackwell, 1989).

27. Strategic trade policy based on rent-shifting arguments has been sharply criticized in an array of analytical and empirical studies as assumption-dependent, reliant on near-perfect information about the actors and the market, and offering relatively small national gains in practice. In addition to those already cited, for examples of recent empirical work, see Kala Krishna, Kathleen Hogan, and Philip Swagel, "The Non-Optimality of Optimal Trade Policies: The U.S. Automobile Industry Revisited, 1979–1985," and Anthony Venables, "Trade Policy Under Imperfect Competition" (Paper presented at Centre for Economic Policy Research and National Bureau of Economic Research conference, in Cambridge, Mass., October 1989).

28. See any previously cited Krugman piece as well as Jeffrey Hart, "Strategic Impacts of High Definition Television for U.S. Manufacturing" (Report for the National Center for Manufacturing Sciences, Ann Arbor, Mich., September 1989), p. 32.

29. For exceptionally useful pieces, see Wilfred Ethier, "Internationally Decreasing Costs and World Trade," *Journal of International Economics*, vol. 9 (1979), pp. 1–24, which extends the work of Jacob Viner, *Studies in the Theory of International Trade* (New York: Augustus Kelley, 1965).

30. Virtuous cycles are associated with the previously discussed positive feedback mechanisms that may translate private increasing returns into a cumulative advantage for the industry or nation. These include interindustry R&D spillovers as well as the knowledge embodied in a skilled work force that is generated by research and production experience that stimulate ongoing productivity improvements in the national economy.

31. In general increasing returns to scale may arise for several reasons. One possibility relates to minimum plant size for efficient production, also known as internal increasing returns. The issue of spillover benefits given in the text refers, however, to another type that is not dependent on the concentration

of production but rather on the size of the downstream watch market worldwide. Wilfred Ethier, "Internationally Decreasing Costs and World Trade," and "National and International Returns to Scale in the Modern Theory of International Trade," *American Economic Review*, vol. 72, no. 3 (June 1982), pp. 389–404.

32. John Zysman and Stephen Cohen, "Double or Nothing: Open Trade and Competitive Industry," *Foreign Affairs*, vol. 61, no. 5 (Summer 1983), pp. 1113–39.

33. Richard Elkus, chairman of the Prometrix Corporation, as reported by the *New York Times*, "The Fast Track to New Markets," May 28, 1989, p. 2.

34. Anne Krueger, "Free Trade is the Best Policy," in Robert Lawrence and Charles L. Schultze, eds., *An American Trade Strategy* (Washington, D.C.: Brookings Institution, 1990). See also Wilfred Ethier, *Modern International Economics* (New York: W. W. Norton, 1983), especially pp. 42–48. For a description of potential gains from trade under dynamic increasing returns, see Krugman, *Rethinking International Trade*, chap. 11.

35. Increased competition and rationalization of industrial structure may increase national purchasing power from trade under imperfect competition, especially under free entry and exit. Unresolved issues in this area are the intertemporal effects on a nation's trade gains from asymmetries in industrial structures and from the exit of competitive domestic firms, and the entry of uncompetitive, subsidized foreign-based firms. See David Richardson, "Empirical Research on Trade Liberalization with Imperfect Competition: A Survey, *OECD Economic Studies*, no. 12 (Spring 1989), pp. 7–50, also as National Bureau of Economic Research Working Paper no. 2883. For a discussion of imperfect competition and welfare effects in the context of Europe's Single Market Program, see Victor Norman, "Assessing Trade and Welfare Effects of Trade Liberalization," *European Economic Review*, vol. 34, no. 4 (June 1990), pp. 725–51.

36. Among other references see Richard Schmalensee and Robert Willig, eds., *Handbook of Industrial Organization* (New York: Elsevier, 1988), pp. 1094–98, and George Eads, "U.S. Government Support of Civilian Technology: Economic Theory versus Political Practice," *Research Policy*, vol. 3 (1974), pp. 5–7.

37. Chapter 5 argues that the HDTV public policy debate in the late 1980s provides a timely illustration of the difficulty, particularly in the midst of the digital electronics revolution, of applying the driver label to a single industry for policy purposes. Research is needed in this area.

38. Industrial policy proponents have argued that the timing of spillovers also matters. Although they may eventually diffuse internationally, they may confer a special advantage to firms in the home country if they are available there first. This first-mover advantage may permit a firm to preempt its rivals from ever gaining a foothold in the industry. Although there is some agreement that the first-mover advantage can play a role in the competitive position of individual firms in some industries, further work still needs to be undertaken to understand the implications for national welfare of the first-mover advantage. A foreign firm, for example, may be the first to develop a productivity-

improving measure, but it may not be appropriate for the domestic environment. Or considerable cost savings in technology development may accrue to the domestic firms in the role as follower firms as the result of the risks taken and the mistakes made by foreign first-mover firms, which were amply demonstrated by the Japanese and Europeans in HDTV. The presence of multinational corporations further complicates the connection between advantages that accrue to particular firms and those that accrue to the nation from first-mover advantage. References regarding these issues include Joseph E. Stiglitz, "Learning to Learn, Localized Learning and Technological Progress," in Partha Dasgupta and Paul Stoneman, eds., *Economic Policy and Technological Performance* (New York: Cambridge University Press, 1987), and Krugman, *Rethinking International Trade*, pp. 165–67.

39. The semiconductor industry, particularly in dynamic random access memories, has been at the forefront, with alliances between U.S. firms and European and Japanese firms dramatically increasing in the 1980s. The list of U.S. firms reads like a Who's Who in high tech, including Texas Instruments, Intel, Advanced Micro Devices, IBM, Motorola, and AT&T.

40. Some have also suggested that knowledge that is local in nature accumulates during the design and production process in firms in the form of skilled workers, proprietary technology, and difficult-to-copy know-how that cannot be purchased or presumably traded through nonmarket mechanisms such as transnational corporate alliances. As discussed later in this chapter, the proliferation of corporate alliances and the internationalization of businesses in the 1980s may have, however, loosened the tie between those benefits local to the firm and the nation's expected return from sponsoring their development through industrial targeting. For references on localized spillovers in production, see Dosi, Tyson, and Zysman, "Trade, Technologies, and Development," especially p. 22.

41. Although not an issue of external economies, a firm's specific knowledge can raise the issue of promotion to capture a larger share of excess returns in an oligopolistic or concentrated industry. Analysis of this idea has weakened rather than strengthened the case for promotion. For critical reviews of the strategic rent-shifting argument, see n. 11.

42. Some have suggested that the strategic industrial policy issue be defined in broad terms such as the promotion of all high-tech industries or the entire electronics sector. Part of the logic behind this suggestion is that the resource diversion costs from promotion would be less because individual industries in the sector that draw on the same resources would not be attracting resources away from each other if they're all targeted. The argument is also a political one. If the target is defined in broad enough terms, fewer will protest and the political-economic costs of the policy will accordingly be less. This compromise overlooks, however, a fundamental point: intervention should try to fix only what is both broken and significant. These critical distinctions are not possible at high levels of aggregation: all high-tech industries. As anecdotal evidence, the disappointing European experience with targeting the entire electronics sector should give policy makers pause before moving toward similar large scale efforts in the United States.

121

43. Krugman, "U.S. Response of Foreign Industrial Targeting," pp. 105–15.

44. On a precautionary note, an emphasis on trade can understate the local nature of some spillovers. Trade may be an imperfect substitute for proximity in some industries. Geographic concentration may facilitate spillovers between industries that are linked by technology, production, or customer relationships. In his study of more than 100 industries in ten countries, Michael Porter found that proximity helped breed effective lines of communication for rapid, ongoing feedback between some upstream and downstream industries. One example often cited is the cluster of interrelated industries that define Silicon Valley. See, for example, Porter, *The Competitive Advantage of Nations* (New York: Free Press, 1990), and Stiglitz, "Learning to Learn, Localized Learning and Technological Progress," pp. 141–43. Others argue that the need for geographic concentration within national borders has diminished with advances in communications and transportation. See, for example, Ethier, "Internationally Decreasing Costs and World Trade."

45. For a basic introduction to this issue, see Ethier, *Modern International Economics*, especially pp. 35–48; for a more detailed analysis, see Ethier, "Internationally Decreasing Costs and World Trade."

46. Krugman, "Strategic Sectors and International Competition," p. 225.

47. Whether the spillovers that have figured prominently in HDTV discussion are national or international in nature has important ramifications for the form of policy intervention. Expanded demand worldwide for semiconductors from the production of HDTV and other high-definition systems (HDTV is one type), for example, can generate cost savings for a variety of industries and nations. Therefore the policy focus should be to open markets further, rather than to stimulate large-scale domestic HDTV production, which could encourage overcrowding and follow-up protectionist measures. By contributing to disputes over market access, industrial policy in this respect seems more likely part of the problem rather than the solution.

48. Eads, "U.S. Government Support for Civilian Technology," p. 15.

49. For a persuasive discussion of the need for, and lack of, well-established rules, see Anne Krueger, an economist who has written widely on the subjects of development and international trade, "Government Failures in Development," *Journal of Economic Perspectives*, vol. 4, no. 3 (Summer 1990), pp. 9–23, and "Free Trade Is the Best Policy," in Lawrence and Schultze, eds., *An American Trade Strategy*, pp. 83–86.

50. For the evidence of clusters in other industries and in other nations, see Porter, *Competitive Advantage of Nations*, table 3.1.

51. Krugman has argued, for example, that "if additional resources of labor and capital are supplied elastically to the industry, the external benefits of larger production will not be confined to the promoting country. Instead they will be passed on to the consumers around the world." See Krugman, "Is Free Trade Passé?" See also Helpman and Krugman, *Market Structure and Foreign Trade*.

52. Porter, *Competitive Advantage of Nations*, pp. 103–21.

53. See also Anne Krueger, "Importance of General Policies to Promote Economic Growth," *World Economy*, vol. 8 (June 1985), pp. 93–108.

54. Laura Tyson, "Managed Trade: Making the Best of the Second Best," in Lawrence and Schultze, *American Trade Strategy*, pp. 160–61.

55. The following discussion draws on a useful review of the strategic industry debate by J. David Richardson, "The Political Economy of Strategic Trade Policy," *International Organization*, vol. 44, no. 1 (Winter 1990), especially pp. 115–23. See also Richard Lipsey and Wendy Dobson, *Shaping Comparative Advantage* (Toronto: C. D. Howe Institute, 1987), especially pp. 114–27.

56. The activist position is reflected in various publications of members of the Berkeley Roundtable on International Economy (BRIE). See, for example, Stephen S. Cohen and John Zysman, *Manufacturing Matters: The Myth of the Post-Industrial Economy* (New York: Basic Books, 1987); Laura Tyson, "Managed Trade: Making the Best of the Second Best," in Lawrence and Schultze, *American Trade Strategy*; and, on semiconductors in particular, Michael Borrus, *Competing for Control: America's Stake in Microelectronics* (Cambridge, Mass.: Ballinger, 1988). On the side of skeptics George C. Eads has taken the position that market failures are minor while American political failures are pervasive and seriously handicap American efforts to manage an economically successful industrial policy; see Eads, "The Political Experience in Allocating Investment: Lessons from the United States and Elsewhere," in Michael Wachter and Susan Wachter, eds., *Toward a New U.S. Industrial Policy?* (Philadelphia: University of Pennsylvania Press, 1983). See also Krueger, "Free Trade Is the Best Policy"; Richardson, "Political Economy of Strategic Trade Policy"; Bhagwati, "Is Free Trade Passé?"

57. Laura Tyson, "Comments on Brander's 'Shaping Comparative Advantage': Creating Advantage, an Industrial Policy Perspective," in Lipsey and Dobson, *Shaping Comparative Advantage*, p. 74.

58. Richardson, "Political Economy of Strategic Trade Policy," pp. 119–22.

59. For critical reviews of industrial policy, see Charles Schultze's influential essay, "Industrial Policy: A Dissent," *Brookings Review*, Fall 1983, pp. 3–12; Philip H. Trezise, "Industrial Policy Is Not the Major Reason for Japan's Success," *Brookings Review*, Spring 1983, pp. 13–18; Robert Lawrence, "Innovation and Trade: Meeting the Foreign Challenge," in Henry Aaron, ed., *Setting National Priorities* (Washington, D.C.: Brookings Institution, 1990); Hugh Patrick, "Japanese Industrial Policy and Its Relevance for United States Industrial Policy," testimony before the Joint Economic Committee of the U.S. Congress, 98th Congress, 1st session, July 13, 1983; George C. Eads, "U.S. Support for Civilian Technology: Economic Theory versus Political Practice," *Research Policy*, Summer 1974, pp. 2–16; Eads, comments at a symposium sponsored by the Federal Reserve Bank of Kansas City, later published in *"Industrial Change and Public Policy,"* August 24–26, 1983 (Kansas City, Mo., pp. 157–68); and Gary Saxonhouse, "What Is All This about 'Industrial Targeting' in Japan?" *World Economy*, vol. 6 (Summer 1983), pp. 253–74. For favorable reviews of industrial policy, see Chalmers Johnson, *MITI and the Japanese Miracle* (Stanford, Calif.: Stanford University Press, 1982), and Chalmers Johnson, Laura D'Andrea Tyson, and John Zysman, *Politics and Productivity: The Real Story of Why Japan Works* (Cambridge, Mass.: Ballinger, 1989).

60. ITC, *Foreign Industrial Targeting* pp. 1–3.

61. In this case industrial policy proponents in Europe and the United States argued that commercial dominance in aircraft would follow from technological superiority and that failure to act would result in a loss of leadership and perhaps irreversible commercial loss. The British and French pursued this objective through the Concorde program, which was started in 1962. They paid little attention, however, to the magnitude and sensitivity of expected demand to price. The result was a commercial flop. Production was terminated in 1979 after only three years of commercial operation. In light of American resistance and the commercial failure of the Concorde despite its technological virtuosity, the United States in this case appeared the winner. For a critical analysis, see George Eads and Richard R. Nelson, "Government Suport of Advanced Civilian Technology: Power Reactors and the Supersonic Transport," *Public Policy*, vol. 19, no. 3 (Summer 1971), pp. 405–28. David C. Mowery and Nathan Rosenberg, "The Commercial Aircraft Industry," in Richard R. Nelson, ed., *Government and Technical Progress: A Cross-Industry Analysis* (New York: Pergamon Press, 1982), pp. 144–46.

62. Daniel Okimoto in his recent book *Between MITI and the Market: Japanese Industrial Policy for High Technology* does not credit industrial policy as the main reason for the relatively high growth rate in Japan: "Rather the secret to Japan's apparent success lies in the overall system within which industrial policy functions." Okimoto further states

> that it is hard to demonstrate beyond a reasonable doubt that the *instrument* of industrial policy has given Japanese companies their competitive edge. Japan's early advances in high technology may have had less to do with the content of industrial policy that with the intrinsic strengths of the private sector, or perhaps the soundness of macroeconomic policies. (p. 230)

He concluded that while some targeted industries succeeded, others generated more headaches than they resolved, and nontargeted sectors such as consumer electronics prospered. On the importance of factors other than industrial policy, see also Hugh Patrick, ed., *Japan's High Technology Industries: Lessons and Limitations of Industrial Policy* (Seattle: Univ. of Washington Press, 1986).

63. Marcus Noland, "Industrial Policy and Japan's Trade Pattern" (Washington, D.C.: Institute for International Economics, June 1990). See also ITC, *Foreign Industrial Targeting*, pp. 1–3.

64. Nelson, *High Technology Policies*.

65. For discussion of the details and effects of the arrangement, see Kenneth Flamm, "Semiconductor," in *Europe 1992: An American Perspective* (Washington, D.C.: Brookings Institution, 1990).

66. As a rationale for discounting this uncertainty and picking industries today, proponents for a more aggressive American policy have argued that the risk and economic costs of not picking will increase over time. But this risk must be balanced against the political difficulties of reversing a poorly chosen path—which will also increase over time.

67. Proposed foreign takeovers of companies supplying Sematech have sparked a debate about the extent of federal responsibility to the consortium and its objectives. See "Drive Opens in Senate to Halt Sale of Semi-Gas," *Electronic News*, October 8, 1990; Senators Lloyd A. Bentsen (D, Tex.) and Jeff Bingaman (D, N.M.), letter to James F. Rill, assistant attorney general, in opposition to sale of Semi-Gas to a subsidiary of Nippon Sanso, September 27, 1990; and Martin Tolchin, "Monsanto Unit Sale Faces Inquiry," *New York Times*, December 21, 1988.

68. Nelson, *High Technology Policies*.

69. Representative Norm Mineta as quoted in "Political Lightning Rod," *Electronics*, October 1990, p. 57. See also the Office of Technology Assessment, *Making Things Better: Competing in Manufacturing* (Washington, D.C.: Government Printing Office, February 1990), p. 211.

70. No one of the activist school has explained either how the logic of their arguments once applied by policy makers would avoid these situations or what would be the economic and policy ramifications. For one attempt at an explanation, see Dosi, Tyson, and Zysman, "Trade, Technologies, and Development," especially pp. 28–29.

71. Dosi, Tyson, and Zysman, "Trade, Technologies, and Development," pp. 24–25.

72. The Aerospace Industries Association listed ten technologies, in *The Leading Edge: Key Technologies for the 1990's* (Washington, D.C., AIA: n.d.) the Computer Systems Policy Project, sixteen, in "Perspectives: Success Factors in Critical Technologies" (Washington, D.C., July 1990); the U.S. Department of Commerce, Technology Administration, listed twelve, in *Emerging Technologies, A Survey of Technical and Economic Opportunities* (Washington, D.C.: Government Printing Office, Spring 1990); the Department of Defense, twenty, in *The Department of Defense Critical Technologies Plan* (Washington, D.C.: Government Printing Office, March 1990); and the Office of Science and Technology Policy, more than twenty technologies in its report. *Report of the National Critical Technologies Panel* (Arlington, Va., March 1991).

73. See press release, "Mineta Introduces Legislation to Strengthen Government Focus on Industries Critical to Future U.S. Economic Growth"; H.R. 1517, 102nd Congress, 1st session; and Council on Competitiveness, *Gaining New Ground: Technology Priorities for America's Future* (Washington, D.C.: 1991).

74. See Office of Technology Assessment, *The Big Picture: HDTV and High Resolution System* (Washington, D.C.: Government Printing Office, June 1990), p. 12. For calls to use the lists of other nations, see Fred Branfman, *Hearings on High Definition Television*; and Lester Thurow as quoted by Jeff Faux in "Whatever Term You Use," p. 23. For differences in national systems of innovations, see Richard R. Nelson and Nathan Rosenberg, *Technical Innovations and National Systems* (forthcoming).

75. Rockwell International's campaign for the B-1 bomber amply illustrates the politics involved with building support for projects. To ensure support from even those policy makers initially opposed to the B-1, a vast complex of

subcontractors was assembled to cover more than 400 of the 435 congressional districts. Under a policy of strategic industry promotion to improve the national economic health, projects may also become dispersed geographically so that policy makers can each have a winning industry or a piece of one in their district. On the B-1, see Hedrick Smith, *The Power Game* (New York: Ballantine Books, 1988), pp. 179–180.

76. Krueger, "Free Trade is the Best Policy," pp. 72, 84–85; Eads, "Government Support for Advanced Technology," p. 14.

77. For a useful discussion of how the implications of advanced-technology industries should be studied, see United States, International Trade Commission, *Identification of U.S. Advanced-Technology Manufacturing Industries*, USITC publication 2319, September 1990.

78. Paul Krugman, "Industrial Policy's Not So Bad," *Journal of Commerce*, May 8, 1990. For a good summary of the managed trade debate, see Lawrence and Schultze, *American Trade Strategy*.

79. Tyson, "Managed Trade."

80. Clyde V. Prestowitz, Jr., "Before We Kill Off Textiles," *Journal of Commerce*, October 3, 1990.

81. Anne Krueger argues in particular that the initial political equilibrium may not be a long-term equilibrium as special interest groups after they receive assistance lobby for more favorable treatment and those not previously favored lobby to protect their interests on the grounds that they are also strategic. See Krueger, "Government Failures in Development." This is not to suggest that the United States should minimize the potential economic importance of emerging, innovative industries. Instead it is to underscore that industrial policy is a deceptively simple approach. The actual policy requires overcoming a complex course of economic, technical, and political hurdles that may not even be worth the effort.

82. The term "industry-led" first became popular in the United States during the HDTV debate. House, Subcommittee on Telecommunications and Finance of the Committee on Energy and Commerce, *Public Policy Implications of Advanced Television Systems*, especially comments of Representative Mel Levine, p. 534, and *Hearings on High Definition Television*, pp. 194–208. See also Rebuild America, "Consortia and Capital."

83. In addition to self-selection, the AEA was also nominated in congressional hearings by other proponents, House, Subcommittee on Telecommunications and Finance of the Committee on Energy and Commerce, *Hearings on High Definition Television*, pp. 201–205.

84. Ibid.

85. Some Europeans have also suggested that the logistics of performing joint research compound the cost problems of advancing civilian technology such that the hassles of joint research may outweigh the benefits. Guy de Jonquieres, "Giving Direction to the Single Market," *Financial Times*, October 22, 1990, and "Shadows over the Sunrise Sector," *Financial Times*, July 2, 1990. For a critical American view of large-scale cooperative research, see Porter, *Competitive Advantage of Nations*.

86. Some have suggested that uncertainty concerning the magnitude of spillovers does not suffice as a case against policy intervention: perhaps the government like business should sometimes take a chance. But political institutions would need to be structured to reflect a business portfolio approach to risk management so that failures are accepted and minimized and that national long-term economic interests are favored rather than those with political power. Until these issues are addressed, the policy guidelines outlined in the text seem appropriate.

Chapter 3: TV and Consumer Electronics

1. The European color television and VCR markets were roughly $9.5 billion (15.9 million units) and $5.1 billion in 1986. Jeffrey Hart, "Strategic Impacts of High-Definition Television for U.S. Manufacturing" (Report prepared for the National Center for Manufacturing Sciences, September 1989, Ann Arbor, Mich.), fig. 4 and pp. 5–7.

2. Jeffrey Hart, "Consumer Electronics," in Bjorn Wellenius, Arnold Miller, Carl Dahlman, and Darius Maas, eds., *Electronics Industry Development* (Washington, D.C.: World Bank, forthcoming), table 4.

3. Jeffrey Hart, "Strategic Impacts of High-Definition Television," tables 1 and 3.

4. U.S. Congressional Budget Office, *Using R&D Consortia for Commercial Innovation: Sematech, X-Ray Lithography, and High Resolution Systems* (Washington, D.C.: Congressional Budget Office, July 1990), p. 54.

5. U.S. Department of Commerce, *1990 U.S. Industrial Outlook* (Washington, D.C.: Government Printing Office, 1990), p. 39-5. For a description of the European marketplace, see Michael Skapinker, "A British Industry Back in the Picture," *Financial Times*, September 26, 1990.

6. Matsushita purchased Motorola's factory in 1974. Philips, a Dutch firm, purchased Magnavox in 1975 and Sylvania from GTE in 1981. Thomson, a French firm, purchased the former RCA consumer electronics division from GE in 1987.

7. This discussion draws on the work of Edward Graham and Paul Krugman, *Foreign Direct Investment in the United States* (Washington, D.C.: Institute for International Economics, 1989), pp. 40–41.

8. Some analysts say that Japanese dumping was the cause. While recognizing this as a factor, others attribute the decline for the most part to wise business and technology strategies of Japanese firms in the 1960s and 1970s and more recently to the global investment strategies of European television manufacturers. For a useful summary of this issue, see Jeffrey Hart, "The Consumer Electronics Industry in the United States: Its Decline and Future Revival," in *The U.S. Electronics Industry Complex* (Report prepared for the U.S. Congress, Office of Technology Assessment, October 1988).

9. "Electronics Industry Association, Consumer Electronics, HDTV, and the Competitiveness of the U.S. Economy," in *Public Policy Implications of Advanced Television Systems* (Staff report for the U.S. Congress, House,

Subcommittee on Telecommunications and Finance of the Committee on Energy and Commerce, 101st Congress, 1st session, February 1, 1989), pp. 379–80. See also Department of Commerce, *1990 Industrial Outlook*, p. 39–7.

10. Other factors that contribute to high domestic content include patents held by U.S.-owned firms on tubes and glass and the post-1985 decline in the dollar relative to the yen, which made manufacturing in the United States more attractive. Television circuitry contributes less than 10 percent to the manufacturing cost and is produced primarily abroad. See Hart, "Strategic Impact of High Definition Television," pp. 4–10, and "The Consumer Electronics Industry in the United States: Its Decline and Future Revival"; testimony of Advanced Television Advisory Committee, Federal Communications Commission, *Public Policy Implications of Advanced Television Systems*, pp. 479–85. See also Robert Nathan Associates, "Television Manufacturing in the United States: Economic Contributions, Past, Present, and Future" (Report prepared for the Electronics Industry Association, February 1989, included in Senate, Committee on Government Affairs, *Hearing on Prospects for Development of A U.S. HDTV Industry*, 101st Congress, 1st session, August 1989), especially pp. 403–407.

11. Committee to Preserve American Color Television (COMPACT) testimony to Subcommittees on Science, Research, and Technology and International Scientific Cooperation of the Committee on Science, Space, and Technology, *Hearing on Should Foreign Companies Participate in Federally Funded Research*, 101st Congress, 1st session, November 1, 1989. See also Hart, "Consumer Electronics."

12. Roughly $70 million has already been invested at Sarnoff for advanced television research. Press release dated January 25, 1990, David Sarnoff Research Center.

13. "Sony Expands in Silicon Valley," *Dataquest Research Bulletin*, May 1989, p. 1. Shin Kusomoki and Masaki Ickikawa, "A Broadcasting Delay," *Electronics*, October 1990, p. 57.

14. Steven Greenhouse, "France's Industrial Shogun," *New York Times*, September 17, 1990, p. 6.

CHAPTER 4: CAUSE FOR ACTIVISM?

1. For proponent arguments see U.S. Congress, House, Subcommittee on Telecommunications and Finance of the Committee on Energy and Commerce, *Public Policy Implications of Advanced Television Systems*, 101st Congress, 1st session, February 1, 1989, and U.S. Congress, House, Subcommittee on Telecommunications and Finance of the Committee on Energy and Commerce, *Hearings on High Definition Television*, 101st Congress, 1st session, March 1989, especially p. 224.

2. Representative Don Ritter, cochairman of the House HDTV Caucus, "To Miss Out on High Definition TV Is to Miss Out on the 21st Century," *Roll Call* Competitiveness Policy Briefing, July 10–16, 1989, pp. 16–17.

3. Dr. Barry Whalen, senior vice president, Microelectronics and Computer Technology Corporation (MCC), in congressional testimony defined high-

definition systems "to include the production, distribution, reception, processing, and display of high quality, high-definition, broadband video and data for consumer, business, scientific, and military applications." U.S. Congress, House, Subcommittee on Telecommunications and Finance of the Committee on Energy and Commerce, *Hearings on Consortia and the Development of High Definition Systems*, 101st Congress, 1st session, September 13, 1989.

4. See, for example, testimony of the American Electronics Association, House, Subcommittee on Telecommunications and Finance of the Committee on Energy and Commerce, *Public Policy Implications of Advanced Television Systems*. See also House, Subcommittee on Telecommunications and Finance of the Committee on Energy and Commerce, *Hearings on High Definition Television*, pp. 199–205.

5. For HDTV activist position, see House, Subcommittee on Telecommunications and Finance of the Committee on Energy and Commerce, *Public Policy Implications of Advanced Television Systems*, especially pp. 22–35, 300.

6. The Japanese commitment to high-definition systems has been estimated between $750 million and $1.5 billion. The Hi-Vision system is part analog and part computer-digital technology: a compromise made on the way to the marketplace that may prove to be costly. For a critical assessment of the Japanese and European HDTV systems, see written testimony of Richard Jay Solomon, Massachusetts Institute of Technology, submitted to U.S. Congress, House, Subcommittee on International Scientific Cooperation of the Committee on Science, Space, and Technology, *Hearings on International Standards and High Definition Television*, 101st Congress, 1st session, May 31, 1989, pp. 70–79.

7. Several unexpected technology and economic developments on the way to the market have, however, recently caused Japan to reassess its approach to HDTV dramatically and to scale back estimates of HDTV market sales.See "HDTV Threat from Abroad?" in chapter 5 and the Appendix.

8. David E. Sanger, "The New T.V. Makes Debut in Japan," *New York Times*, December 6, 1990, p. D1. Also Michael Skapinker, "Battle at the Television's Sharp End," *Financial Times*, special section on Japan, December 3, 1990, p. 8.

9. Jack Robertson, "High Definition Look at the Future," *Electronic News*, December 3, 1990.

10. This discussion of Japanese HDTV is based on Jeffrey Hart, "U.S. Group Studies High Definition Television in Japan" (Summary of trip to Japan, December 4, 1989), and Peter Fletcher, "Gathering Steam," *Electronics*, October 1990, pp. 48–51.

11. "High Definition Television: The World at War," *Economist*, August 4, 1990, p. 58.

12. The decision of British Satellite Broadcasting to merge with Sky Television and broadcast in PAL (the conventional broadcasting system in the United Kingdom) instead of HD-MAC has raised new questions about the prospects of the European MAC approach. See, for example, Michael Skapinker, "Satellite Groups' Merger Clouds HDTV Picture," *Financial Times*, December 3, 1990.

13. As quoted by Fletcher, "Gathering Steam," p. 48.

14. The Japanese and European systems are limited regarding the future

exchange of video information with advanced computers because their systems are analog and use interlaced scanning while computers are digital and use progressive scanning.

15. House, Subcommittee on Telecommunications and Finance of the Committee on Energy and Commerce, *Hearings on High Definition Television*, p. 200.

16. Ibid., pp. 223, 257; House, Subcommittee on Telecommunications and Finance of the Committee on Energy and Commerce, *Public Policy Implications of Advanced Television Systems*, p. 34.

17. For references to this argument and the struggles of congressman to uderstand it, see U.S. Senate, Subcommittee on Science, Space, and Technology of the Committee on Commerce, Science, and Transportation, *Hearings on High Definition Television*, 101st Congress, 1st session, May 16, 1989, especially the comments of Sen. Albert Gore (D, Tenn.), pp. 140–43; U.S. Senate, Committee on Governmental Affairs, *Hearings on Prospects for Development of U.S. HDTV Industry*, 101st Congress, 1st session, August 1, 1989.

18. Referring to color television, see Susan Walsh Sanderson, *The Consumer Electronics Industry and the Future of American Manufacturing* (Washington, D.C.: Economic Policy Institute, 1989), p. 23.

19. Jeffrey Hart and Laura Tyson, "Responding to the Challenge of HDTV," *California Management Review*, Summer 1989, pp. 132–45; Stephen Cohen and John Zysman, *Manufacturing Matters: The Myth of Post-Industrial Economy* (New York: Basic Books, 1987). For an opposing view of the link between manufacturing and human capital development, see David Richardson, "The Political Economy of Strategic Trade Policy," *International Organization*, vol. 44 (Winter 1990), pp. 107–135.

20. Ritter, "To Miss Out on High Definition TV," p. 16.

21. HDTV sets will require more video memory, faster digital signal and image processors, and more complicated analog/digital hybrid circuits than conventional (NTSC) receivers. There is some controversy over how much memory HDTV will require. Testimony of the Electronics Industry Association to the U.S. Congress, Senate, Committee on Government Affairs, *Prospects for Development of a U.S. HDTV Industry*, 101st Congress, 1st session, August 1, 1989, p. 155.

22. U.S. Office of Technology Assessment, *Making Things Better: Competing in Manufacturing* (Washington, D.C.: Government Printing Office, March 1990), p. 86.

23. Bandwidth influences the amount of information that can move through the system. Accordingly it has been a major challenge with which engineers have been forced to contend in developing HDTV and other high-definition systems that will use digital video.

24. Fifteen or more Japanese companies have reportedly invested an estimated $690 million in 1990 in the development of active matrix displays.

25. For comparative analysis see U.S. International Trade Commission, *High Information Flat Panel Displays and Subassemblies Thereof from Japan*, USITC Publication 2311, September 1990, pp. A-14, A-15. For a discussion of non-CRT display technologies and other industry interests, see Jeffrey Hart,

"Strategic Impact of High-Definition Television" (Report to the National Center for Manufacturing Sciences, September 1989), pp. 18–22.

26. Samuel Weber, "It's a New Age for LCDs," *Electronics*, May 1989, pp. 96–97.

27. Regardless of which technologies prove to be commercially viable, some U.S. industry analysts believe that high-resolution displays will be the semiconductors of the twenty-first century and will have a pervasive influence on other electronic applications and the national economy. For a useful overview of the status of American firms in display technologies, see John Carey, Neil Gross, and Robert Hof, "Bigger, Wider, Flatter, Brighter: The Key to Making It in HDTV," *Business Week*, February 26, 1990, pp. 82–85; David E. Mentley, "The Flat Panel Display Industry in the United States: An Overview" (Paper presented to the U.S. Department of Commerce, Washington, D.C., February 15, 1990); and ITC, *High Information Content Flat Panel Displays*.

28. Wesley R. Iversen, "U.S. Gropes for Unity on HDTV," *Electronics*, March 1989, p. 75, and Hart and Tyson, "Responding to the Challenge of HDTV."

29. Organization for Economic Co-operation and Development, *The Telecommunications Industry: The Challenges of Structural Change* (Paris: OECD, 1988), and Jonathan David Aronson and Peter F. Cowhey, *When Countries Talk: International Trade in Telecommunications Services* (Washington, D.C.: American Enterprise Institute, 1988).

30. See any of the congressional hearings already cited as well as Hart and Tyson, "Responding to the Challenge of HDTV," p. 139.

31. For a skeptical view of the relative importance of residential demand or consumer video services, see House, Subcommittee on Telecommunications and Finance of the Committee on Energy and Commerce, *Public Policy Implications of Advanced Television Systems*, p. 179.

32. Hart and Tyson, "Responding to the Challenge of HDTV," p. 139.

33. See, for example, U.S. General Accounting Office, *High Definition Television: Applications for this New Technology* (Washington, D.C.: U.S. Government Printing Office, December 1989), pp. 6–7.

34. In the 1980s there has been a great deal of discussion of the increasingly interdependent nature of civilian and military technologies and businesses that in some cases makes it impossible to separate military security interests from national interests in a strong and growing economy. Much work still needs to be done in understanding how the two sectors of the economy will interact in the 1990s and what the implications will be for policy. In the meantime, if the real objective is to improve national economic welfare, policy makers should be wary of national security arguments for strategic industries that seek to bypass the economic criteria.

35. House, Subcommittee on Telecommunications and Finance of the Committee on Energy and Commerce, *Public Policy Implications of Advanced Television Systems*, pp. 19, 534.

36. Fred Branfman before House, Subcommittee on Telecommunications and Finance of the Committee on Energy and Commerce, *Hearings on High Definition Television*, p. 205.

37. American Electronics Association, *Development of a U.S.-Based ATV Industry* (Washington, D.C.: AEA, May 9, 1989). For more on motivations and objective of industry-led plan to revitalize consumer electronics, see testimony of Pat Hill Hubbard, AEA, in House, Subcommittee on Science, Technology, and Space, of the Committee on Commerce, Science, and Transportation, *Hearings On High Definition Television*, 101st Congress, 1st session, May 16, 1989, pp. 42–46. For an opposing perspective on the need for the ATV Corporation from other industry representatives, see Donald Johnstone, president and chief operating officer, Philips Consumer Electronics, in *Hearings on High Definition Television*, especially pp. 55–56.

38. House, Subcommittee on Telecommunications and Finance of the Committee on Energy and Commerce, *Public Policy Implications of Advanced Television Systems*, especially p. 41.

39. After the initial proposal was poorly received by the Bush administration and other policy makers, CECC was renamed the Electronics Capital Corporation. As of mid-1990 the details of this plan were not publicly available. For CECC, see National Advisory Committee on Semiconductors, ''A Strategic Industry at Risk'' (Report to the president and Congress, Washington, D.C., November 1989).

40. National Advisory Committee on Semiconductors, *A Strategic Industry at Risk* (Arlington, Va.: NACS, November 1989), especially pp. 2–3, 23.

41. The bills in the 101st Congress were H.R. 1267, H.R. 1516, H.R. 2287, S. 952, and S. 1001. Hearings were held, but none of the bills progressed beyond the committee stage, and all died in the 101st Congress.

42. Representative Richard Gephart (D, Mo.) led the charge for the Democrats as the anchorman for the Democratic action agenda. Others included Senator John Glenn (D, Ohio) with plan for a civilian Defense Advanced Research Project Agency (DARPA), Senator Albert Gore (D, Tenn.) on HDTV, Representative Ernest F. Hollings (D, S.C.) with promotion of the Advanced Technology Program in the Commerce Department, Senator Jeff Bingaman (D, N.M.) in forming critical technology lists, and Representative Edward J. Markey (D, Mass.) on HDTV.

43. Larry F. Darby, Darby Associates, *Economic Potential of Advanced Television Products* (Report for the National Telecommunications and Information Administration, U.S. Department of Commerce, Washington, D.C., April 7, 1988); Robert R. Nathan Associates, Industry Research and Analysis Group, *Television Manufacturing in the United States: Economic Contributions—Past, Present, and Future* (Report prepared for the Electronics Industries Association, Washington, D.C., November 22, 1988), chap. 4, ''High Definition T.V.'s Potential Economic Impact on Television Manufacturing in the United States''; and American Electronics Association, Advanced Television Task Force, Economic Impact Team, *High Definition Television (HDTV): Economic Analysis of Impact* (Santa Clara, Calif.: AEA, November 1988).

44. For critical reviews of these market forecasts see Shawn O'Donnell, ''Forecasting the HDTV Market,'' MIT Media Laboratory, submitted to House, Subcommittee on Telecommunications and Finance of the Committee on Energy and Commerce, *Public Policy Implications of Advanced Television Systems*;

and Philip Webre, *The Scope of the High-Definition Television Market and its Implications for Competitiveness* (Washington, D.C.: Congressional Budget Office, July 1989).

45. American Electronics Association, ATV Task Force Economic Impact Team, *High Definition Television: Economic Analysis of Impact* (Working document, Santa Clara, Calif., November 1988), pp. 3-3–3-12. By AEA calculations control of 50 percent of American HDTV, VCR, and TV sales by U.S.-owned firms would generate an additional $42.5 billion in personal computer sales for U.S.-owned PC firms over the weak HDTV scenario in which U.S. firms control less than 10 percent of the U.S. HDTV market. Similarly a strong position of U.S.-owned HDTV firms as compared with a weak position is expected to generate an additional $3.5 billion of automated manufacturing equipment in the year 2000 for U.S.-owned firms.

46. U.S. Congress, Senate, Committee on Governmental Affairs, *Prospects for Development of a U.S. HDTV Industry*, 101st Congress, 1st session, August 1, 1989, p. 29.

47. *HDTV Newsletter*, January 1989, p. 10.

48. For a skeptical review of the AEA study in particular, see the report of the FCC Advisory Committee on Advanced Television Service, House, Subcommittee on Telecommunications and Finance of the Committee on Energy and Commerce, *Public Policy Implications of Advanced Television Systems*, especially pp. 472–80.

49. Hart and Tyson, "Responding to the Challenge of HDTV," p. 144.

CHAPTER 5: CAUSE FOR SKEPTICISM

1. Representative Don Ritter, cochairman of the House HDTV Caucus, "To Miss Out on High Definition TV Is to Miss Out on the 21st Century," *Roll Call*, "Competitiveness Policy Briefing," July 10–16, 1989, pp. 16–17.)

2. U.S. Federal Communications Commission, Advisory Committee on Advanced Television, in U.S. Congress, House, Subcommittee on Telecommunications and Finance of the Committee on Energy and Commerce, *Public Policy Implications of Advanced Television Systems*, 101st Congress, 1st session, March 1989, p. 478.

3. The greater sophistication of a HDTV set relative to a conventional set will make its initial price relatively high. An optimistic scenario with early models of 40-inch, or larger, inexpensive high-resolution displays puts the retail price of an HDTV receiver at $1,500 versus a unit price of more than $3,000 in the pessimistic scenario with no breakthroughs in developing and manufacturing large inexpensive displays. As production experience is acquired and demand increases, prices would fall to some extent. Jeffrey Hart has estimated that the circuitry cost, which contributes roughly 11 percent to set cost, would decline from roughly $400 to $50–80 per set as component firms acquired production experience. See Hart, "Strategic Impacts of High-Definition Television for U.S. Manufacturing" (Report to the National Center for Manufacturing Sciences, Ann Arbor, Mich., September 1989), pp. 16–22.

4. Preliminary market studies suggest that if sets are 28 inches or smaller, consumers can detect a difference but are not willing to pay the HDTV premium price. A MIT study suggests that consumers prefer HDTV but are not willing to pay more than a $100 premium to get it. The study was limited by the technology available at the time in its capability to test consumer willingness to pay for HDTV. Both 18-inch and 28-inch CRTs were used to display HDTV images in the study rather than a 40-inch that many believe is needed to distinguish HDTV and conventional television. See W. Russel Neuman's study, "The Mass Audience Looks at HDTV: An Early Experiment" (Paper presented to the annual convention of the National Association of Broadcasters, Las Vegas, April 1988). See also the written testimony of Richard J. Solomon, to Senate, Committee on Government Affairs, *Hearings on Prospects for Development of a HDTV Industry*, p. 3. For a sample of questions raised about the nature and degree of potential HDTV demand, see House, Subcommittee on Telecommunications and Finance of the Committee on Energy and Commerce, *Public Policy Implications of Advanced Television Systems*, particularly pp. 280–81 and 547–48.

5. Widespread delivery of HDTV will require studios to convert existing equipment to the HDTV standard. NBC cost estimates of this process for local stations ranges from $4 million to $40 million. The HDTV conversion cost of the NBC network, excluding expenses of owned stations, would exceed $500 million. As a result of these costs, NBC has proposed an intermediate step of EDTV before introducing full HDTV. See the NBC press release, March 30, 1990, and Shandle, "Just How Much Will HDTV Conversion Cost?" p. 66.

6. If HDTV is introduced in the form of fully interactive TV sets capable of exchanging information with home computers, then the service value of HDTV and the potential demand would tend to increase, as would that of products emerging from the computer industry.

7. In 1990, 1,446 TV stations operated in the United States with another 200 or so construction permits outstanding. See Jack Shandle, "Just How Much Will HDTV Conversion Cost?" *Electronics*, June 1990, p. 67, and U.S. Congress, Office of Technology Assessment, *The Big Picture: HDTV and High Resolution Systems* (Washington, D.C.: Government Printing Office, June 1990).

8. Edmund L. Andrews, "Advanced T.V. Testing Set amid Tumult on Technology," *New York Times*, November 15, 1990.

9. Among other references see Joyce Anne Oliver, "Consumer Electronic Trends: High Interest in High-Definition," *New Technology*, April 1990, for a description of some of the applications of high-definition technology already emerging in industrial workstations, medical imaging, and auto design that are expected to precede consumer demand for HDTV.

10. "Core" refers here to the digital data processing, transmission, display, and manufacturing technologies associated with information systems. U.S. Congress, Office of Technology Assessment, *Making Things Better* (Washington, D.C.: Government Printing Office, February 1990), pp. 85, 88.

11. George Gilder, "Forget HDTV, It's Already Outmoded," *New York Times*, May 28, 1989, p. III-2; "IBM-TV?" *Forbes*, February 20, 1989, pp. 72–76. For a view of the future that endorses an analog-based HDTV (like that developed

in Japan) as a necessary intermediate step toward Gilder's version of the future, see Jeffrey Hart and Laura Tyson, "Responding to the Challenge of HDTV," *California Management Review*, vol. 31, no. 4 (Summer 1989), p. 140.

12. Similar to the comments of HDTV proponents, Hart and Tyson, "Responding to the Challenge of HDTV," p. 140, argue that "it will be very difficult, if not impossible, to develop a digital TV industry without the technological and manufacturing base that will be developed for HDTV." HDTV needs to be taken as a necessary intermediate step for the future described by Gilder.

13. The Office of Technology Assessment argues that although the super-computer accounts for a small share of the total electronics market, it is still important in driving a number of the underlying technologies. But this argument does not apply to consumer HDTV, for which even the OTA report emphasizes the large potential consumer market as the source of productivity improvements in the underlying technologies. See OTA, *The Big Picture: HDTV and High Resolution Systems*, especially pp. 11, 61, 65–66, 76.

14. U.S. Congressional Budget Office, *The Scope of High-Definition Television Market and Its Implications for Competitiveness* (Washington, D.C.: CBO, July 1989).

15. "High Definition Television: The World at War," *Economist*, August 4, 1990, p. 59. See also Congressional Budget Office, *The Scope of High-Definition Television*, pp. 21–23.

16. For the semiconductor case, see the National Advisory Committee on Semiconductors, *A Strategic Industry at Risk* (Washington, D.C.: NACS, 1989)

17. In separate reports Kenneth Flamm of the Brookings Institution and Philip Webre of the Congressional Budget Office, for example, found that HDTV sets are likely to be less important than computers as a source of chip demand, with HDTV likely to account for less than 1 percent of the global chip market. CBO, *The Scope of High-Definition Television Market and Its Implications for Competitiveness*, pp. 21–23; Kenneth Flamm comments before U.S. Congress, Senate, Subcommittee on Science, Technology, and Space of the Committee on Commerce, Science, and Transportation, *Hearings on High Definition Television*, 101st Congress, 1st session, May 16, 1989, pp. 97, 140–42.

18. U.S. Congress, Senate, Committee on Government Affairs, *Hearing on Prospects for Development of a U.S. HDTV Industry*, 101st Congress, 1st session, August 1, 1989, pp. 497–506.

19. For references to two reports on the computer industry as the locomotive in display, see Jack Robertson on a forthcoming report from the Defense Science Board's High-Resolution Task Force. Based on a prepublication copy, Robertson reports that the task force believes the world computer industry would be as important a technology and market driver for large high-definition display as the frequently cited consumer HDTV product. "DOD Panel Urges $500 million Flat Panel Display Program," *Electronic News*, July 2, 1990, p. 31; Congressional Budget Office, *The Scope of the High Definition Television Market*.

20. Joseph Donahue of Thomson Consumer Electronics, who has worked in flat panel research for more than thirty years, does not view flat panels as a viable medium for HDTV in the foreseeable future. Others apparently share

135

his skepticism as eight companies recently invested $1 billion in a more conventional display medium (tubes and glass). Donahue's comments on a working copy of this study, August 17, 1990; Peter F. McCloskey, EIA, in *Hearings on Prospects for the Development of a U.S. HDTV Industry*, p. 154.

21. In terms of static resolution, computer monitors are more sophisticated than television monitors, but the display of full-motion video continues to be a problem on these monitors. Richard Jay Solomon, Program on Communications Policy and the Media Laboratory, Massachusetts Institute of Technology, testimony before U.S. Congress, House, Subcommittee on International Scientific Cooperation of the Committee on Science, Space, and Technology, *Hearings on Impact of International Standards on High Definition Television*, 101st Congress, 1st session, May 31, 1989, p. 6. Hart, "Strategic Implications of High-Definition Television"; Clive Cookson, "First Auditions for the Multimedia Show," *Financial Times*, October 11, 1990.

22. Donahue, comments on this study, August 17, 1990.

23. Among others Kenneth Flamm has emphasized the competitive and regulatory framework as an important factor in determining rates of technological and market convergence. See Robert Crandall and Kenneth Flamm, eds., *Changing the Rules: Technological Change, International Competition and Regulation in Communications* (Washington, D.C.: Brookings Institution, 1989), especially pp. 1–61.

24. James C. McKinney, testimony to the House, Subcommittee on International Scientific Cooperation of the Committee on Science, Space, and Technology, *Hearings on Impact of International Standards on High Definition Television*, annex B, p. 3.

25. Based on Dataquest data as cited by Bernard Levine, "PC, Workstation Firms Face Off," *Electronics*, June 26, 1989, p. 30.

26. Bernard Levine, "PC, Workstation Firms Face Off," pp. 29–30. See also Lawrence Curran, "It's Becoming a Buyer's Market," *Electronics*, January 1990, pp. 57–59; U.S. Department of Commerce, International Trade Administration, *U.S. Electronics Sector* (Washington, D.C.: Government Printing Office, April 1990), p. 107.

27. Peter T. McCloskey, president, EIA, in Senate, Committee on Government Affairs, *Hearings on Prospects for Development of a U.S. HDTV Industry*, pp. 155–156.

28. International Trade Administration, *U.S. Electronics Sector*, pp. 107–10.

29. Sidney Topol, letter to the editor, "The Rocky Road to HDTV," *New York Times*, July 29, 1990; Edmund L. Andrews, "The Lure of Digital Television," *New York Times*, June 29, 1990. See also George Eads and Richard R. Nelson, "Government Support of Advanced Technology: Power Reactors and the Supersonic Transport," *Public Policy*, vol. 19 (Summer 1971), pp. 405–419, for the British experience with the Concorde and how the pursuit of technological success led to commercial failure.

30. The United States accounted for roughly 20 percent of the global computer market or $62 billion in 1989. In 1988 U.S. computer manufacturers accounted for 17 percent of total U.S. industry R&D or $10 billion; "R&D Scoreboard," *Business Week*, special innovation ed., 1989. The following section

draws on Jack Shandle, "Who Will Dominate the Desktop in the '90's?" *Electronics*, February 1990, pp. 48–50; Clive Cookson, "First Auditions for the Multimedia Show," *Financial Times*, October 11, 1990; and "An IBM Tagalong Sets Independent Course, with Plenty of Risks," *Wall Street Journal*, April 21, 1989.

31. Personal computers are the largest slice of the computer market with $17.6 billion in 1988 sales and expected growth of 8 percent in 1989. HDTV sets and VCRs are not projected to reach this level of sales until sometime after 2005. "1989 U.S. and Overseas Market Forecast," *Electronics*, January 1989, pp. 59–69.

32. "A Video Window of Opportunity Opens," *Financial Times*, July 7, 1989. See also "A New Picture for Computer Graphics," *Wall Street Journal*, July 5, 1989.

33. The work is also based on a contract under the High-Definition Display System Project of the Defense Agency for Advanced Research Projects (DARPA), Department of Defense.

34. Attention has been sparked by the prospect of a 24-hertz frame rate production standard with progressive scanning that would allow an easy conversion to 72-hertz rate of workstation monitors. For the Zenith proposal, see Jack Shandle, "HDTV: Looking Good as a CAD/CAM Standard," *Electronics*, June 1990, pp. 70–71. For recent moves in the private sector to increase cooperation between TV and computer manufacturers, see Richard Doherty, "Tapping Digital Video," *Electronic Engineering Times*, December 3, 1990.

35. Crandall and Flamm, *Changing the Rules*; Hart and Tyson, "Responding to the Challenge of HDTV."

36. Although outnumbered by HDTV activists at a congressional hearing, Kenneth Flamm tried to channel the HDTV policy debate away from an industrial policy perspective toward infrastructure issues. See his statement to the U.S. Congress, House, Subcommittee on Telecommunications and Finance of the Committee on Energy and Commerce, *Hearings on High Definition Television*, 101st Congress, 1st session, March 9, 1989, pp. 239–40. For an overview of communication issues see G. A. Keyworth, "Goodbye, Central: Telecommunications and Computing in the 1990's" (Hudson Institute, Indianapolis, March 26, 1990).

37. David E. Sanger, "The New T.V. Makes Debut in Japan," *New York Times*, December 6, 1990, p. D1.

38. See Mark Nelson and Bob Hagerty, "EC Stumbles on a Standard for HDTV," *Wall Street Journal*, November, 1990; "High Definition Television: On the Blink," *Economist*, February 23, 1991, p. 66.

39. For an interesting analysis of the motivations by stage of the product cycle for corporate alliances and nonprice competition, see Ashoka Mody, *Staying in the Loop: Alliances for Shaping Technology*, Discussion paper 61 (World Bank, Washington, D.C., 1989).

40. The importance of globalization varies across industries. The creative labeling of "research" by both American- and foreign-owned firms has, for example, limited agreement among analysts. For the positive interpretation

see the National Academy of Engineering, *National Interests in a Global Economy* (Washington, D.C.: National Academy Press, 1991). For a skeptical review see Laura D'Andrea Tyson, testimony to U.S. Congress, Joint Economic Committee, *Hearings on National Interests in an Age of Global Industry*, 101st Congress, 2nd session, September 13, 1990.

41. Anne Brunsdale, "Global Competitiveness in the Computer Industry" (Speech given to the Computer and Communications Industry Association, Washington, D.C., May 23, 1989).

42. Trade permits a larger downstream market to develop and provides an opportunity for greater specialization of labor and product differentiation to generate international, industry-level scale economies. These economies do not require geographic concentration, which, if countries have similar cost advantages, encourages intra-industry trade or the sharing of industries. On this subject the work of Wilfred Ethier is an exceptionally clear and useful guide. See, for example, "Internationally Decreasing Costs and World Trade," *Journal of International Economics*, vol. 9 (1979), pp. 1–24, and "National and International Returns to Scale in the Modern Theory of International Trade," *American Economic Review*, vol. 72 (June 1982). See also discussion on international scale economies in "Challenging Tradition" section in chapter 2 of this volume.

43. In addition to the strengths of Intel and Texas Instruments and recent work of AMD in microprocessors, the David Sarnoff Research Center in Princeton, New Jersey, has dramatically cut the development time for high-definition systems and other signal-processing research and development projects that depend on digital signal-processing algorithms. One estimate is that the design time for an image-processing chip could be cut by a factor of ten with a reduction in cost by a factor of five. See Jack Shandle, "Real-Time Simulation Speeds HDTV Designs," *Electronics*, March 1990, pp. 68–69, and "Motherboard on a Chip Becomes a Reality," *Electronics*, October 1990, p. 77.

44. This reference treats the European Community as a single market. The Commerce Department has reported that a range of new competitors has emerged to challenge the U.S. position in some computer technologies but also concluded that U.S. strength in advanced computers is unlikely to be undermined. The rapid pace of U.S. product introductions and shortened product life cycles will make it difficult for the Japanese or others to reduce the two-generation lead of U.S. firms. International Trade Administration, *Competitive Status of the U.S. Electronics Sector*, pp. 11, 15, 43, and 112–20. On American software strengths see also comments of Tom Friel, vice president, Consumer Electronics Group, Electronics Industry Association, *HDTV Newsletter*, January 1989, p. 4.

45. Information provided to the author from the SIA vice president, Warren Davis, October 2, 1990. For initial announcement see David E. Sanger, "Industries in U.S. and Japan Form Alliance on New T.V. Technology," *New York Times*, November 9, 1989.

46. Jack Robertson, "Groups Reach Accord on FMVs," *Electronic News*, October 8, 1990, p. 8.

47. For two useful articles see Charles Leadbeater, "Japan Gets to the Heart of ICL," *Financial Times*, September 31, 1990, and John Burgess and Evelyn Richards, "Does Foreign Investment in U.S. Pose a Threat?" *Washington Post*, October 23, 1990.

48. Donald Johnstone, president and chief executive officer, Philips Consumer Electronics, House, *Hearings on High Definition Television*, pp. 56, 92.

49. Electronics Industry Association, in House, *Hearings on Prospects for Development of a U.S. HDTV Industry*, pp. 30–31.

50. The American Electronics Association eventually conceded that U.S. location of HDTV production would not dramatically alter the current competitive situation of U.S. chip firms. Senate Committee on Government Affairs, *Hearings on Prospects for Development of a U.S. HDTV Industry*, p. 43. See also Hart, "Strategic Impact of High-Definition Television," pp. 41, 35 n. 54, who notes that U.S.-owned semiconductor producers made the decision not to supply chips for consumer electronics applications even when U.S.-owned consumer electronics firms were still dominant. Thus it is far from clear that ownership of downstream products such as consumer HDTV would stimulate sales of U.S.-owned semiconductors.

51. Hart, "Strategic Impacts of High Definition Television for U.S. Manufacturing," p. 41.

52. Marc Levison, "A Plot Too Complex for TV," *Journal of Commerce*, October 15, 1990.

53. A central problem is determining the nature of the Japanese pricing strategy. Is it predatory, or is it a wise business strategy based on forward pricing? The latter recognizes the characteristics of the industry in which production experience is paid for by low initial yields and rewarded over time with falling average costs as yields increase. Additionally, more than 70 percent of Japanese high-information-content flat panels are sold in markets other than the United States: in other words the relative strength of Japanese firms may not be dependent on dumping and predatory pricing in the U.S. market. International Trade Commission, *High Information Content Flat Panel Displays and Subassemblies Thereof from Japan*, USITC publication 2311 (Washington, D.C., September 1990).

54. For a sample of the different views see the list of HDTV policy agendas submitted to House, Subcommittee on Telecommunications and Finance of the Committee on Energy and Commerce, *Public Policy Implications of Advanced Television Systems*.

55. Jagdish Bhagwati, "Is Free Trade Passé after All?" *Weltwirtschaftliches Archiv*, vol. 125, no. 3 (1989), pp. 19–20, 28.

56. Robert Noyce, president, Sematech, quoted in "Pentagon Aim to Revive U.S. T.V. Industry," *Washington Post*, December 19, 1989, and his interview with Lee Smith, "How the U.S. Can Compete Globally," *Fortune*, June 5, 1989, p. 248. Pat Choate, vice president of policy analysis, TRW, has suggested that required purchases from U.S.-owned semiconductor producers and other forms of local-content restrictions would be the least-expensive option for the government to stimulate the creation of a domestic industry. See his testimony

in addition to those of Fred Branfman before House, Subcommittee on Telecommunications and Finance of the Committee on Energy and Commerce, *Hearings on High Definition Television*, pp. 189–90, 208–209. See also testimony of the American Electronics Association, House, Subcommittee on Telecommunications and Finance of the Committee on Energy and Commerce, *Public Policy Implications of Advanced Television Systems*, pp. 39–41, and Craig Fields on the need for government procurement to guarantee initial market for domestic HDTV producers at Computer Business Equipment Manufacturers Association conference, Washington, D.C., March 1989.

Chapter 6: Policy Implications

1. For a critical review of the Japanese and European systems, see Richard Jay Solomon, testimony, in U.S. Congress, House, Subcommittee on International Scientific Cooperation of the Committee on Science, Space, and Technology, *Hearings on High Definition Television: The International HDTV Standard Setting Process*, 101st Congress, 1st session, May 31, 1989, pp. 73–74.

2. In particular the rallying cry at congressional hearings was that "the United States needs to gain an industry [HDTV] to retain an industry [electronics]." See Pat Hubbard, for the American Electronics Association, in U.S. Congress, Senate, Committee on Government Affairs, *Hearings on Prospects for Development of a U.S. HDTV Industry*, 101st Congress, 1st session, August 1, 1989, p. 29. AEA has recently backed off from this position as it became clear that the HDTV set and consumer electronics market will not be the critical one for semiconductors or flat panels. This retrenchment is a particularly revealing illustration of the difficulty, discussed in chapter 2, involved today with picking an industry on which economic progress and national competitiveness will depend.

3. U.S. Congress, Senate, Subcommittee on Defense Industry and Technology of the Committee on Armed Services, *The Future of the U.S. Semiconductor Industry and the Impact on Defense*, 101st Congress, 1st session, November 29, 1989, p. 82; also definitions provided to the author by the Office of Science and Technology Policy in November 1990 as extension of the report "U.S. Technology Policy," Executive Office of the President, September 26, 1990.

4. It is difficult to determine precisely the extent of federal funding because advanced imaging technologies have an array of applications and because program budgets have not had a separate line item for the individual technologies. As a result, research in advanced imaging may be embodied in the development of a computer system sponsored by the government but may not have been counted as part of the $120 million estimate. Office of the President, *Budget of the United States Government, Fiscal Year 1991* (Washington, D.C.: Government Printing Office), p. 85. Also author's telephone communication with Nancy Schwartz in the Office of Management and Budget, February 1990.

5. According to DARPA's high-definition display technology program manager, Marco Slusarczuk, the budget for the high-definition system

program increased to $74.5 million for fiscal year 1991. Telephone communication, January 1991.

6. Defense Advanced Research Projects Agency request for proposals, December 23, 1988.

7. For the list of R&D projects for equipment, see Congressional Budget Office, *Using R&D Consortia for Commercial Innovation: Sematech, X-ray Lithography, and High Resolution Systems* (Washington, D.C.: CBO, July 1990), especially pp. 23–26.

8. DARPA moved under the direction of Craig Fields in early 1990 to serve as a venture capitalist to boost commercial competitiveness in particular areas raised, for example, fundamental questions that have yet to be resolved about who should be responsible for addressing whatever capital market imperfections may exist in the development of dual-use technologies and products. On HDTV and the $4 million DARPA investment in Gazelle Microelectronics Inc. that ignited a debate over the venture capital issue and that contributed to Field's resignation (removal) see Evelyn Richards, "Pentagon Aim to Revive U.S. T.V. Industry," *Washington Post*, December 19, 1988; Evelyn Richards, "Doubting the Focus of HDTV," *Washington Post*, May 31, 1989, p. H1. Andrew Pollack, "Silicon Valley Investment by Pentagon," *New York Times*, April 10, 1990.)

Theodore H. Moran, Georgetown University, has written a useful paper on the implications of a global defense industrial base. He asks in particular when this trend should be worrisome, embraced, or ignored. Moran suggests that market concentration, rather than foreign ownership or dependency on foreign supplies, is the condition against which the military needs to guard. See "Globalization of America's Defense Industries: Managing the Threat of Foreign Dependence," *International Security*, (Summer 1990), pp. 57–99.

9. Successful development of high-definition systems and the need to improve manufacturing technology may indicate a government role at the state or national level in helping to close the technology gap that may arise between small and large firms in this area. If the diffusion of best manufacturing practices proves to be a problem, then the existing regional industrial extensions programs of the National Institute of Standards and Technology may be a form of support that should be evaluated.

10. David Mowery comments on working copy of this study. See also the National Academy of Engineering, *National Interests in a Global Economy* (Washington, D.C.: National Academy Press, 1991).

11. See Advisory Council on Federal Participation in Sematech, "Sematech 1990" (Report to Congress, May 1990), especially pp. ES-2, ES-3, for an example of how the spread of international networks has blurred the meaning of national technological capability. See Charles Leadbeater, "Japan Gets to the Heart of ICL," *Financial Times*, September 31, 1990. On Sematech, the many references to the U.S.-Japan alliances include Jacob M. Schlesinger, "AT&T, NEC Agree to Cooperate," *Wall Street Journal*, April 23, 1991, p. B4.

12. For a summary of prospective benefits and costs of consortiums, see testimony of Claude E. Barfield, American Enterprise Institute, "Industrial Consortia," in U.S. Congress, House, Subcommittee on Science, Research,

and Technology of the Committee on Science, Space, and Technology, *Hearings on Government Role in Joint Production Ventures*, 101st Congress, 1st session, September 12, 1989.

13. A precautionary note should be added that past government-to-government plans for large-scale collaboration have exposed a tendency to introduce noneconomic factors that limit the potential cost effectiveness of this form of collaboration. Private participants would be wise to avoid projects like the European Aircraft Project in which four nations collaborated to produce an advanced military fighter but at the cost of escalating expenses. Among other references, see Steven Greenhouse, "European Fighters: Cost U.S. Pride," *New York Times*, December 21, 1989, p. D-1.

14. Charles Leadbeater, "EC Urges 4 States to Stop Industrial Subsidies," *Financial Times*, July 20, 1990, and Lucy Kellaway, "Brussels Flexes Its Muscles to Take a Swing at State Subsidies," *Financial Times*, November 20, 1989, p. 6.

15. The U.S.-Japan Structural Impediment Initiative is one approach to pulling domestic policies into international negotiations. For reasons to reject this approach, see Rudiger Dornbush, "The Structural Impediments Initiative Talks Are a Joke," *International Economy*, February-March, 1990. See also Robert Z. Lawrence and Charles L. Schultze, eds., *An American Trade Strategy: Options for the 1990s* (Washington, D.C.: Brookings Institution, 1990), and Sylvia Ostry, *Governments and Corporations in a Shrinking World*.

16. The complex question of where and how implementation of competition policies should be harmonized is relevant because asymmetries in national enforcement (or toleration) can function as a barrier to foreign market entry, which may impair the efficiency of U.S.-based operations, may distort production location decisions, and may ultimately undercut support for a rules-based international trading system. For a preliminary study of some of these issues, see Robert Z. Lawrence, "Do Keiretsu Reduce Japanese Imports?" (Paper prepared for the Japanese Economic Planning Symposium, Tokyo, January 22–23, 1991).

17. The rising importance of multinationals in domestic, as well as international, affairs may limit the tendency toward nationalistic policies. Questions have been raised about the enthusiasm of MNCs for their role of sustaining a multilateral system based on accepted rules. For an overview of this issue and new trade policy challenges, see Sylvia Ostry, *Governments and Corporations in a Shrinking World* (New York: Council on Foreign Relations Press, 1990).

Gertrude Himmelfarb
Distinguished Professor of History
  Emeritus
City University of New York

Samuel P. Huntington
Eaton Professor of the
  Science of Government
Harvard University

D. Gale Johnson
Eliakim Hastings Moore
  Distinguished Service Professor
  of Economics Emeritus
University of Chicago

William M. Landes
Clifton R. Musser Professor of
  Economics
University of Chicago Law School

Sam Peltzman
Sears Roebuck Professor of Economics
  and Financial Services
University of Chicago
  Graduate School of Business

Nelson W. Polsby
Professor of Political Science
University of California at Berkeley

Murray L. Weidenbaum
Mallinckrodt Distinguished
  University Professor
Washington University

Research Staff

Claude E. Barfield
Resident Scholar

Walter Berns
Adjunct Scholar

Douglas J. Besharov
Resident Scholar

Robert H. Bork
John M. Olin Scholar in Legal Studies

Anthony R. Dolan
Visiting Fellow

Dinesh D'Souza
Research Fellow

Nicholas N. Eberstadt
Visiting Scholar

Mark Falcoff
Resident Scholar

Gerald R. Ford
Distinguished Fellow

Murray F. Foss
Visiting Scholar

Suzanne Garment
DeWitt Wallace Fellow in
  Communications in a Free Society

Patrick Glynn
Resident Scholar

Robert A. Goldwin
Resident Scholar

Gottfried Haberler
Resident Scholar

Robert W. Hahn
Resident Scholar

Robert B. Helms
Visiting Scholar

Charles R. Hulten
Visiting Scholar

Karlyn H. Keene
Resident Fellow; Editor,
  The American Enterprise

Jeane J. Kirkpatrick
Senior Fellow

Marvin H. Kosters
Resident Scholar; Director,
  Economic Policy Studies

Irving Kristol
John M. Olin Distinguished Fellow

Michael A. Ledeen
Resident Scholar

Robert A. Licht
Resident Scholar

Chong-Pin Lin
Associate Director, China Studies
  Program

John H. Makin
Resident Scholar

Allan H. Meltzer
Visiting Scholar

Joshua Muravchik
Resident Scholar

Charles Murray
Bradley Fellow

Michael Novak
George F. Jewett Scholar;
  Director, Social and
  Political Studies

Norman J. Ornstein
Resident Scholar

Richard N. Perle
Resident Fellow

Thomas W. Robinson
Director, China Studies Program

William Schneider
Resident Fellow

Bernard Schriever
Visiting Fellow

Herbert Stein
Senior Fellow

Irwin M. Stelzer
Resident Fellow

Edward Styles
Director, Publications

W. Allen Wallis
Resident Scholar

Ben J. Wattenberg
Senior Fellow

Carolyn L. Weaver
Resident Scholar

A Note on the Book

This book was edited by Ann Petty of the
publications staff of the American Enterprise Institute.
The text was set in Palatino, a typeface designed by
the twentieth-century Swiss designer Hermann Zapf.
Scott Photographics, Inc., of Riverdale, Maryland,
set the type, and Edwards Brothers Incorporated,
of Ann Arbor, Michigan, printed and bound the book,
using permanent acid-free paper.

The AEI PRESS is the publisher for the American Enterprise Institute for Public Policy Research, 1150 17th Street, N.W., Washington, D.C. 20036: *Christopher C. DeMuth,* publisher; *Edward Styles,* director; *Dana Lane,* assistant director; *Ann Petty,* editor; *Cheryl Weissman,* editor; *Susan Moran,* editorial assistant (rights and permissions). Books published by the AEI PRESS are distributed by arrangement with the University Press of America, 4720 Boston Way, Lanham, Md. 20706.